WiMax Explained

Dr. Kalai Kalaichelvan
Lawrence Harte

ALTHOS

Althos Publishing
Fuquay-Varina, NC 27526 USA
Telephone: 1-800-227-9681
Fax: 1-919-557-2261
email: info@althos.com
web: www.Althos.com

Althos

Copyright 2007 By Althos Publishing
First Printing

Printed and Bound by Lightning Source, TN.

International Standard Book Number: 1-932813-54-3

About the Authors

Dr. Kalai Kalaichelvan

Dr. Kalai Kalaichelvan holds a Ph.D. and M.Sc. from the University of Toronto, Toronto, Canada. His comprehensive knowledge and unique perspective have propelled him to lead Canada's telecommunications industry. Since founding the company in 2001, Dr. Kalaichelvan has grown EION into a leader in broadband wireless networking earning a place on the Deloitte Wireless Fast 50 list and a Product of the Year Award for its WiMAX products. Dr. Kalaichelvan's gifted knowledge of the telecom industry has led him to speaking engagements at the World Bank and the title Innovator of the Year from the National Research Council and the CATA Alliance.

Lawrence Harte

Mr. Harte has over 29 years of technology analysis, development, implementation, and business management experience. Mr. Harte has worked for leading companies including Ericsson/General Electric, Audiovox/Toshiba and Westinghouse and has consulted for hundreds of other companies. He has authored over 80 books on telecommunications technologies and business systems covering topics including IPTV, mobile telephone systems, data communications, voice over data networks, broadband, prepaid services, billing systems, sales, and Internet marketing. Mr. Harte holds many degrees and certificates including an Executive MBA from Wake Forest University (1995) and a BSET from the University of the State of New York, (1990).

WiMax Explained

Table of Contents

What is WiMax

Worldwide Interoperability for Microwave Access (WiMAX) is a wireless communication system that allows computers and workstations to connect to high-speed data networks (such as the Internet) using radio waves as the transmission medium with data transmission rates that can exceed 120 Mbps for each radio channel [1]. The WiMAX system is defined in a group of IEEE 802.16 industry standards and its various revisions are used for particular forms of fixed and mobile broadband wireless access.

WiMAX is primarily used as a wireless metropolitan area network (WMAN). Derived from wireless metropolitan area networks (WMAN), WiMAX can provide broadband data communication access services ranging from urban to rural settings.

Used throughout the world, WiMAX broadband competes with DSO, cable modem and optical broadband connections by offering applications which include consumer broadband wireless Internet services, interconnecting lines (leased lines), and transport of digital television (IPTV) services.

Figure 1.1 depicts a number of the applications compatible/suitable for the wireless WiMAX systems including broadband Internet access, telephone access services, television service access and mobile telephone services.

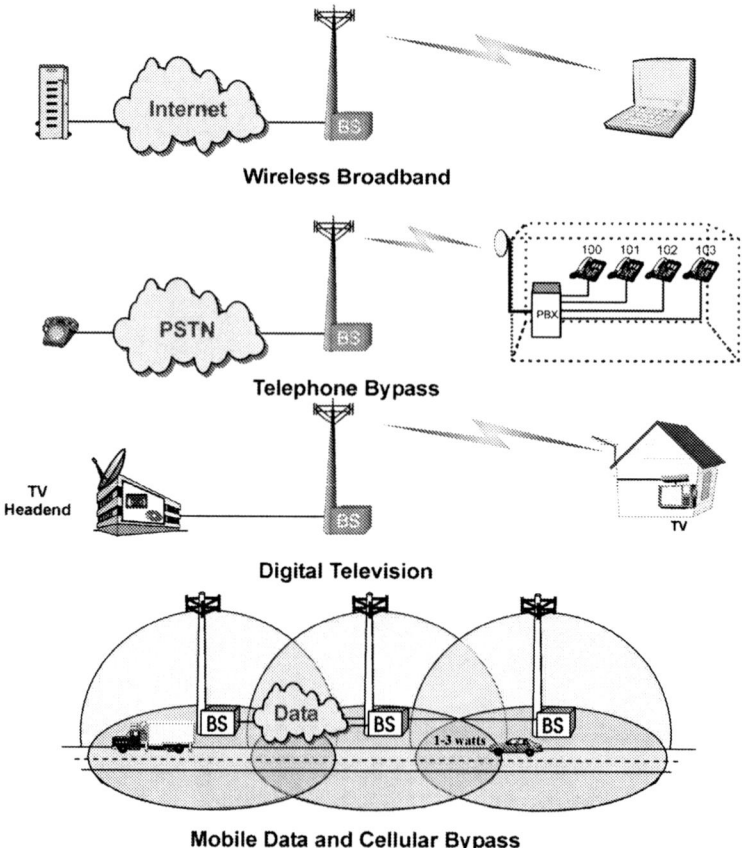

Figure 1.1, WiMax Applications

The 802.16 system was initially designed for fixed location nomadic service in order to provide communication services to more than one location. While nomadic service may be provided to many locations, it typically requires the transportable communication device to be in a fixed location during the usage of communication services.

Developed for mobile service, the 802.16e specification adds mobility management, extensible authentication protocol (EAP), handoff (call transfer), and power saving sleep modes.

WiMAX has several different physical radio transmission options which allows it to be deployed in areas with different regulatory and frequency availability requirements. Moreover, the system was designed with the ability to be used in licensed or unlicensed frequency bands using narrow or wide frequency channels.

Figure 1.2 illustrates a variety of uses that WiMAX networks can provide including point-to-point links, residential broadband and high-speed business connections. As shown, the point to point (PTP) connection may be independent from all other systems or networks. The point to multipoint (PMP) system allows a radio system to provide services to multiple users. WiMAX systems can also be established as mesh networks allowing the WiMAX system to forward packets between base stations and subscribers without having to install communication lines between base stations.

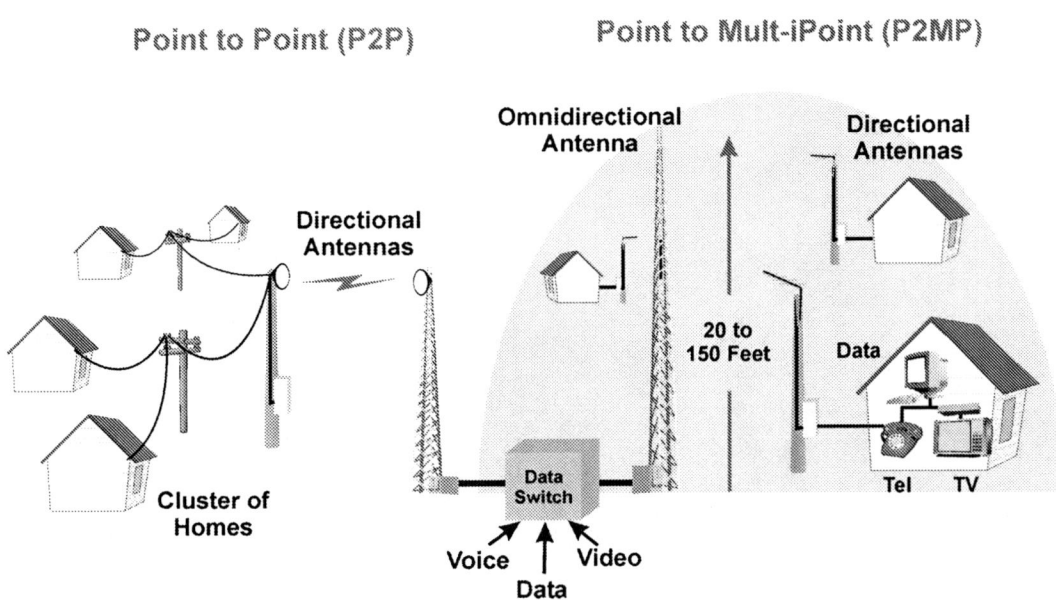

Figure 1.2, Types of WiMAX Systems

WiMAX systems are composed of subscriber stations, base stations, interconnecting switches, and databases. Subscriber stations receive and convert radio signals into user information, while base stations are the radio part of a radio transmission site (cell site). Base stations convert signals from subscriber stations into a form that can be transferred into the wireless network. Interconnecting switches and transmission lines transfer signals between base stations and other systems (such as the public telephone network or the Internet). Databases are collections of data that is interrelated and stored in memory (disk, computer, or other data storage medium). WiMAX systems typically contain several databases that hold subscriber information, equipment configuration, feature lists and security codes.

Figure 1.3 illustrates the key components of a WiMAX radio system. The major components of a WiMAX system include; a subscriber station (SS), a base station (BS) and interconnection gateways to datacom (e.g. Internet), telecom (e.g. PSTN) and television (e.g. IPTV). An antenna and receiver (subscriber station) in the home or business converts the high frequency

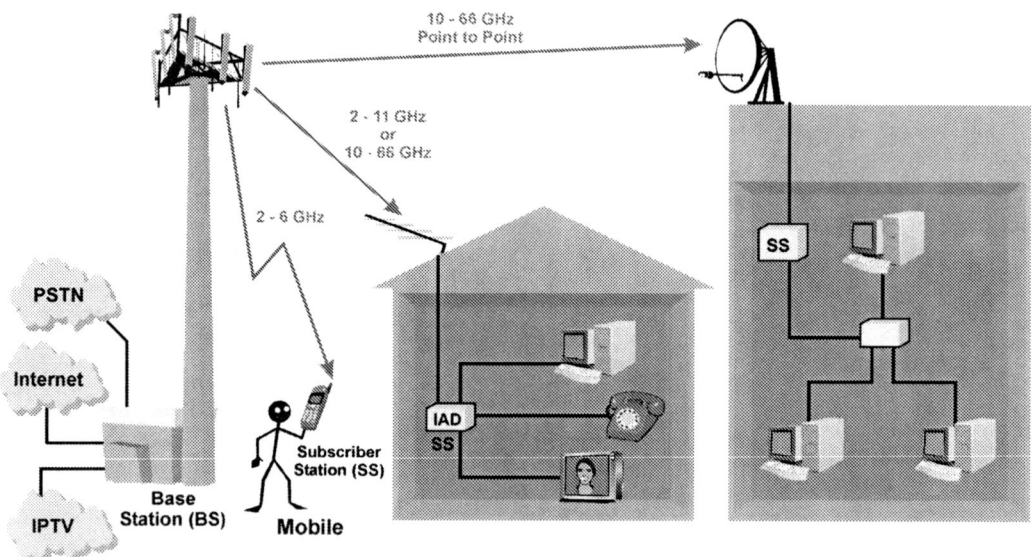

Figure 1.3, WiMax Radio System

(microwave) radio signals into broadband data signals for distribution. In figure 1.3 a WiMAX system is being used to provide television and broadband data communication services. When used for television services, the WiMAX system converts broadcast signals to a data format (such as IPTV) for distribution to IP set top boxes. When WiMAX is used for broadband data, the WiMAX system also connects the Internet through a gateway to the Internet. This example also shows that the WiMAX system can reach distances of up to 50 km for fixed point to point operation.

To develop a cost effective, high-speed data transmission WMAN system, the IEEE created the 802.16 industry specification. The original 802.16 systems were a line of sight system that operates in the 10 GHz to 66 GHz of radio spectrum. To allow the 802.16 systems to operate in the 2 GHz to 11 GHz bands, the 802.16A specification was created.

The radio channel bandwidth of a WiMAX system can be very wide (e.g. greater than 20 MHz) and the radio access technology uses dynamically assigned burst transmission. This allows WiMAX systems to provide data transmission rates that can exceed 120 Mbps [2].

To help ensure WiMAX products perform correctly and are interoperable with each other, the WiMAX Forum was created. The WiMAX Forum is a non-profit organization that certifies products conform to the industry specification and interoperate with each other. WiMAX™ is a registered trademark of the WiMAX Alliance and the indication that the product is WiMAX Certified™ indicates products have been tested and should be interoperable with other products regardless of who manufactured the product.

Because the fundamental technology used in the 802.16 system is similar to 802.11 (wireless LAN), which is similar to 802.3 (Ethernet), wired or wireless LANs systems can be connected to a WiMAX system as an extension. In some cases, the WiMAX system can be operated independently to provide direct data connections between all the computers that can connect to the WiMAX system.

Advantages of WiMAX

Advantages of WiMAX include the use of a standardized technology, rapid deployment, spectral efficiency, penetrating radio coverage, scalability, security, high data throughout, quality of service, and cost effectiveness. WiMAX is an industry-standardized technology. This allows multiple manufacturers to produce compatible equipment, which usually results in lower cost equipment.

WiMAX systems can be rapidly deployed. Because each WiMAX transmitter site can serve hundreds of square kilometers of area, it is possible to deploy high-speed WiMAX communication services in a city or relatively large geographic region with weeks or months.

The technologies used in WiMAX enable it to be very spectrally efficient. Spectral efficiency is a measurement characterizing a particular modulation and coding method that describes how much information can be transferred in a given bandwidth. This is often given as bits per second per Hertz. The WiMAX system uses very efficient modulation and coding methods to achieve spectral efficiency that is higher than mobile telephone or other types of wireless systems.

WiMAX can use a mix of robust types of access technologies. This enables WiMAX systems to provide radio coverage into a wide range of geographic areas. This allows WiMAX signals to penetrate through and around trees and into buildings when necessary.

WiMAX systems are scalable. It has the ability to increase the number of users or amount of services it can provide without significant changes to the hardware or technology used. The WiMAX system can be expanded through the addition of radio channels, transmitter sites and smart antenna systems giving it virtually unlimited scalability.

There are various security processes that are available in WiMAX systems. Security is the ability of a system or service to maintain its desired well being or operation without damage, theft or compromise of its resources from unwanted people or events. The WiMAX system has a security later integrated into its overall operation permitting reliable authentication and encryption of user and system data. High security also permits revenue assurance and strong privacy which can increase consumer confidence. High security also offers the ability of WiMAX system operators to entice high-value content owners to allow distribution of their content (e.g. movie programs).

WiMAX systems have the potential to provide very high data transmission rates. Data throughput is the amount of data information that can be transferred through a communication channel or transfer through a point on a communication system. The WiMAX system has the capability of using wide channel bandwidths and segmenting of data rates providing a very high potential data throughput rate. WiMAX systems offer a much higher potential data transmission rate than almost all other wireless systems giving system operators a competitive advantage.

WiMAX systems can be configured to offer services that have different types of quality of service (QoS) levels. QoS is one or more measurement of desired performance and priorities of a communications system. QoS measures may include service availability, maximum bit error rate (BER), minimum committed bit rate (CBR) and other measurements that are used to ensure quality communications service. The QoS capabilities of WiMAX systems permit system operators to provide priority services to high-value customers and best effort services to less demanding consumers.

The costs of providing WiMAX data communication services (cost effectiveness) can be lower than other types of wireless systems. The highly efficiently modulation and coding characteristics of WiMAX systems enables the providing services to more customers per radio channel than alternative types of systems. This means that WiMAX can have lower capital cost and operational cost per customer.

Figure 1.4 shows some of the key features and benefits offered by WiMAX systems. WiMAX features include cost effective service, ability to offer different levels of QoS, high data transmission rate capability, various types of security, system scalability, efficient radio utilization and the ability to rapidly setup and deploy standard WiMAX equipment.

Feature	Benefit
Standard Technology	Lower cost equipment.
Rapid Deployment	Rapid recovery of investment.
Spectral Efficiency	Fewer radio channels required for each user.
Radio Coverage	Good radio coverage through trees and in buildings.
Scalability	Ability to expand as customer demand increases. Pay as you grow.
Security	Revenue assurance, consumer confidence and access to premium content sources.
High Data Throughput	Competitive advantage.
Quality of Service	High reliability and guaranteed services to high value customers.
Cost Effectiveness	Lower capital and operational cost per customer.

Figure 1.4, WiMax Key Features and Benefits

WiMAX Compared to 802.11 Wi-Fi

802.11 Wi-Fi is an industry standard developed by the IEEE for wireless network communication to provide wireless local area network (WLAN) services. It usually operates in the 2.4 GHz or 5.8 GHz spectrum and permits data transmission speeds from 1 Mbps to 54 Mbps.

WiMAX differs from Wi-Fi in various ways including service range, data transmission throughput, quality of service capability and security processes.

Outlined in figure 1.5, 802.16 WiMAX and 802.11 Wi-Fi systems are designed to provide different capabilities. WiMAX systems are well suited for wide area networks that are managed by a service provider and Wi-Fi systems are a good choice for local area networks.

Characteristic	802.11 (Wi-Fi)	802.16 (WiMax)
Range	Typically up to 100 meters	Up to 50 km range.
Coverage	Designed for indoor transmission	Optimized for more variable outdoor transmission
Scalability	Single cell and fixed channel bandwidth	Frequency reuse and bandwidth range of 1.5 MHz to 20 MHz
Bit Rate	54 Mbps per channel (2.7 bps/Hz)	Over 100 Mbps per channel (5.0+ bps/Hz)
Quality of Service (QoS)	Decentralized QoS control	Coordinated QoS control

Figure 1.5, Differences between WiMAX and Wi-Fi

WiMAX Compared to Mobile Telephone Data Systems

Mobile telephone systems are fully automatic wide-area high-capacity RF networks made up of a group of coverage sites called cells. As a subscriber passes from cell to cell, a series of handoffs ensures smooth call continuity.

Mobile telephone systems have evolved to offer a mix of voice and packet data services. These systems are composed of interlinked cells that have the capability to transfer connections from tower to tower. The radio channel bandwidth is relatively narrow compared to WiMAX systems and the modulation types are less efficient (i.e. more robust). Therefore, the maximum data rates of mobile telephone data systems are lower than that of WiMAX.

Figure 1.6 shows how WiMAX is positioned to fit with cellular data and Wi-Fi systems. WiMAX systems are designed to provide centrally managed high-speed data services over wide areas, whereas Wi-Fi systems are designed to provide self-managed wireless data services over relatively small geographic areas.

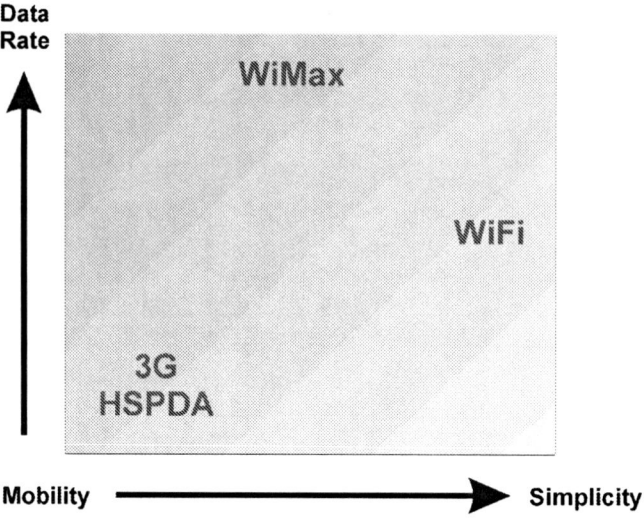

Figure 1.6, Comparison between WiMAX and 3G

Finally, mobile telephone data services are designed to provide a mix of voice and medium speed data services to customers as they move throughout a mobile system.

Data Transmission Rates

Data transmission rate refers to the amount of digital information that is transferred over a transmission medium over a specific period of time and is commonly measured in the amount of bits that are transferred per second (e.g. bps, Mbps).

The data transmission rate for WiMAX systems varies based on factors including radio channel bandwidth (1.25 MHz to 28 MHz), modulation type (BPSK, QPSK, QAM) and channel coding type (percentage of bits dedicated to control and error protection). The raw data transmission rate of a WiMAX radio channel can be in excess of 155 Mbps using QAM modulation. The allocated data transmission rates to each user are typically 1 to 3 Mbps allowing WiMAX operators to have several hundred subscribers for each RF channel.

WiMAX Service Rates

A rate plan is the structure of service fees that a user will pay for services. Rate plans are usually divided into monthly and usage fees.

In the early 2000s, wide area wireless systems (e.g. mobile telephone systems) were limited to relatively low data transmission rates (regularly below 20 kbps). Users were often required to pay for bandwidth on a time or usage basis. Consequently, the cost of this data transmission was approximately 10 cents per kilobyte ($100 per megabyte).

The ability of WiMAX systems to provide high data transmission rates with a limited amount of equipment allows for a substantial reduction in operation cost which thereby leads to significant lowering of service rates. The service rate cost for WiMAX systems in 2006 was approximately 2 to 3 cents per megabyte (99.97% lower than mobile data rates in the mid-1990s).

Figure 1.7 shows a sample wireless broadband service rate plan for a specific WiMAX system in 2006. This example shows that users may select from 3 rate plan options; basic, standard or premium. Each rate plan has a maximum downlink and uplink data transfer rate associated with it along with a monthly service fee. The amount of allocated data that can be transferred as part of the rate plan (at no cost) may be limited (e.g. 100 MByte maximum) or it may be unlimited (e.g. premium plan). If the user exceeds the allocated data transfer amount for their selected rate plan, they are charged a usage fee (e.g. 2 to 3 cents per MByte). In addition to the service fees, WiMAX service providers may charge a connection fee (setup fee) and there may be a leasing charge for equipment that is provided by the WiMAX service provider (e.g. WiMAX modem).

	Basic	Standard	Premium
Downlink Data Transfer Rate	768 kbps	1.5 Mbps	3.0 Mbps
Uplink Data Transfer Rate	256 kbps	768 kbps	1.5 Mbps
Monthly Fee	$19.99	$29.99	$49.99
Monthly Data Transfer	100 MB	500 MB	Unlimited
Cost per MByte	3 cents	2 cents	N/A
Monthly Modem Lease	$10	$10	$10
Connection Fee	$50.00	$50.00	$50.00

Figure 1.7, Sample Wireless Broadband Service Rate Plan 2006

Radio Coverage Area

Radio coverage occurs when a geographic area receives a radio signal above a specified minimum level. WiMAX can operate up to 50 km under line of sight (LOS) and up to 8 km under non-line of sight (NLOS) conditions [3]. Practical cell sizes are limited to approximately 5 miles [4].

WiMAX radio coverage varies based on the options installed and used (such as diversity transmission) in the equipment and the modulation (such as QAM –vs- QPSK), frequency and the parameters that are set.

For the most part, there is a tradeoff between data transmission rate and distance. As the modulation type becomes more efficient (more bits per Hertz), the higher the channel quality has to be at the receiver which means the maximum distance that can be used from the transmitter is reduced.

Radio signal attenuation varies from approximately 20 dB per decade in free space to between 40 to 60 dB per decade when signals travel through objects (resulting in building penetration loss). As the distance increases by a factor of ten in freespace, the signal level drops by a factor of 1000, whereas when radio signals travel through objects (walls and floors), the signal may decrease by a factor of 100,000 or more.

Figure 1.8 illustrates the maximum distance and data transmission rates for fixed and mobile WiMAX communication in a geographic setting. A 20 MHz wide WiMAX radio channel can provide approximately 75 Mbps of data transfer (when it is close to the base station) while the data transmission rate decreases as the distance from the base station increases.

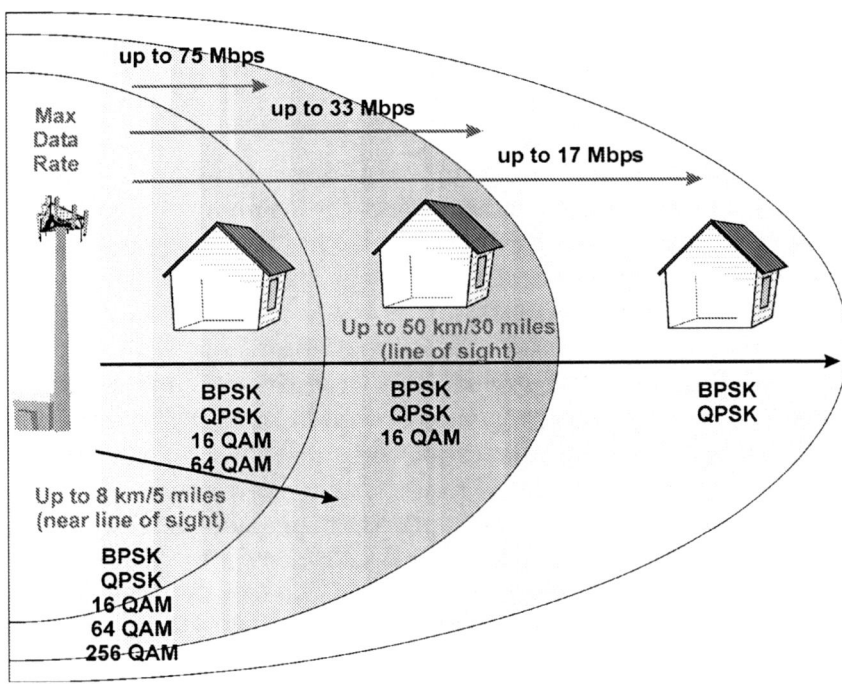

Figure 1.8, WiMax Carrier Serving Area

Frequency Bands

Frequency bands are the range of frequencies that are used or allocated for radio services. There are two primary frequency bands defined for WiMAX systems; 10 to 66 GHz (the original frequency band) and 2 to 11 GHz. The WiMAX system is designed to allow operation on licensed or unlicensed radio channels.

A licensed frequency band is a range of frequencies that requires authorization for use (a license) from a regulatory agency or owner of the frequency band in a geographic area for permission to transmit radio signals in that area. Unlicensed frequency bands are a range of frequencies that can be used by any product or person provided the transmission conforms to transmission characteristics defined by the appropriate regulatory agency.

Channel Loading

Channel loading is a ratio of the number of users authorized to operate on a particular channel or systems compared to the number of users that actively transmit on a system. An example of channel loading on a WiMAX system is the number of broadband subscribers that can be effectively served by a single WiMAX radio channel.

The amount of channel loading depends on a variety of factors including the type of use (e.g. bursty web browsing or watching continuous digital telephony). For many types of applications, the subscriber station does not usually continuously transmit data while it is connected to the WiMAX system. For Internet browsing, the typical data transmission activity is less than 10%. This could allow channel loading of 10:1 or more. For example, in a WiMAX cell or single radio coverage area that has a channel capacity of 70 Mbps, 700 broadband Internet customers could be provided with data rates of 1 Mbps each.

The service provider can affect the channel loading through their price plans. Price plans can range from usage based service (charge per megabyte transferred) to unlimited rate plans. In 2006, the services and rate plans offered by WiMAX service providers were similar to digital subscriber line (DSL) and cable modem services rate plans.

Spectral Efficiency

Spectral efficiency is a measurement characterizing a particular modulation and coding method that describes how much information can be transferred in a given bandwidth. This is often given as bits per second per Hertz.

Modulation and coding methods that have high spectral efficiency are typically very sensitive to small amounts of noise and interference and often have low geographic spectral efficiency. WiMAX was designed to use multiple types of modulation, which allows the system to offer very high spectrum efficiency when the signal quality permits. Because of the robust modulation type used for the GSM system, its spectral efficiency is approximately 1.0 to 1.5 bits per Hertz. More efficient modulation types are used in the WCDMA system providing 1.5 to 2.5 bits per Hertz. 802.11 Wi-Fi systems can use efficient modulation types when channel quality is acceptable (e.g. limited interference by other unlicensed devices), which can provide spectral efficiency of 2 to 3 bits per Hertz. The 802.16 WiMAX system can use very efficient modulation providing an efficiency of 3 to 5 bits per Hertz.

Figure 1.9 depicts the approximate spectral efficiency for several different types of systems. This diagram shows that the spectral efficiency of early mobile telephone systems (e.g. GSM) is approximately 1.0 to 1.5 bits per Hertz and that newer cellular systems (such as WCDMA) have spectral efficiencies of 1.5 to 2.5 bits per Hertz. The spectral efficiency of 802.11 WLAN system can be 2 to 3 bits per Hertz and the spectral efficiency of the 802.16 WiMAX system is approximately 3 to 4 bits per Hertz.

Fixed WiMAX

Fixed wireless is the use of wireless technology to provide voice, data, or video service to fixed locations. Fixed wireless services include wireless local loop (WLL), point-to-point microwave, wireless broadband, and free-space optical

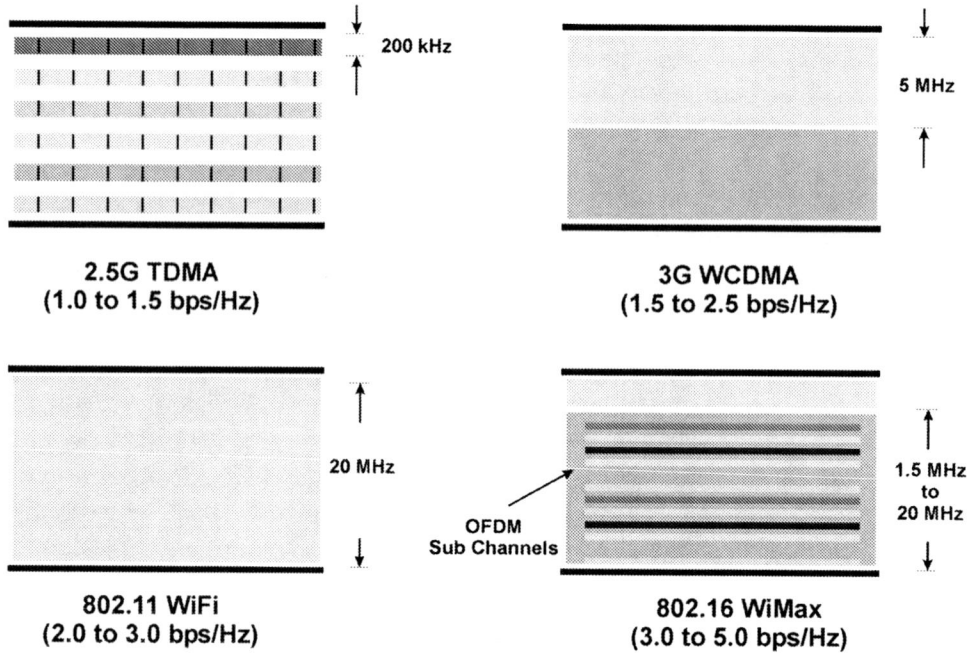

2.5G TDMA
(1.0 to 1.5 bps/Hz)

3G WCDMA
(1.5 to 2.5 bps/Hz)

802.11 WiFi
(2.0 to 3.0 bps/Hz)

802.16 WiMax
(3.0 to 5.0 bps/Hz)

Figure 1.9, WiMax Spectral Efficiency

communication. Fixed wireless systems may replace or bypass wired telephone service, high-speed telephone communication links, and cable television systems.

The initial WiMAX 802.16 standard was developed to provide high-speed data
communication for licensed fixed applications at microwave frequencies (10-
66 GHz). Shortly after the development of the initial 802.16 standard started,
several versions were created to provide different service types and to operate
in lower frequencies (2-11 GHz).

The 802.16a specification was created to allow WiMAX to operate in the 2-11
GHz range. This was followed by the 802.16c specification, which contained

profiles for 10-66 GHz systems. Development of an 802.16d specification was started to define profiles for 2-11 GHz range. Eventually, all of the variations (802.16a, 802.16c and 802.16d) were merged into a single 802.16 specification (802.16-2004).

Figure 1.10 illustrates that point to point communication can reach up to approximately 50 km (30 miles) with a data rate of over 72 Mbps. For multi-point communication that can operate in partially blocked (near line of sight) terrain, the WiMAX system can reach up to 10 km (6 miles) with a data rate of over 40 Mbps.

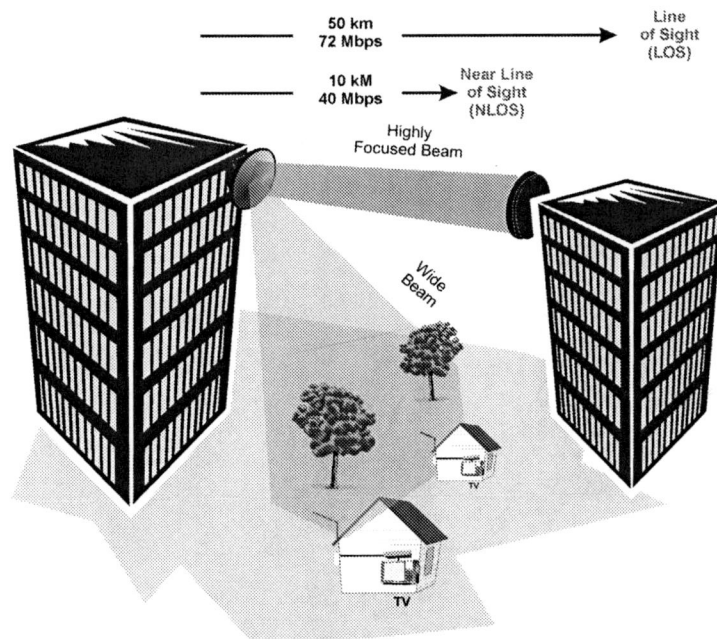

Figure 1.10, WiMax Fixed Communication Services

Mobile WiMAX

Mobile wireless is the use of wireless technology to provide voice, data or video service to locations that may change over time. Mobile wireless services include mobile telephone, mobile data and broadcast video services. Mobile wireless systems may replace or augment wireless telephone service, high-speed data communication links, and broadcast television systems.

Figure 1.11 depicts how a WiMAX system provides communication services to mobile devices. The WiMAX system has various options that can improve the signal quality to mobile users, which operate in a variety of urban, pedestrian and rural environments. The WiMAX system can use OFDM modulation which enhances mobile performance in multipath signal conditions and it can use MIMO and beam forming capabilities to increase signal reliability and system capacity.

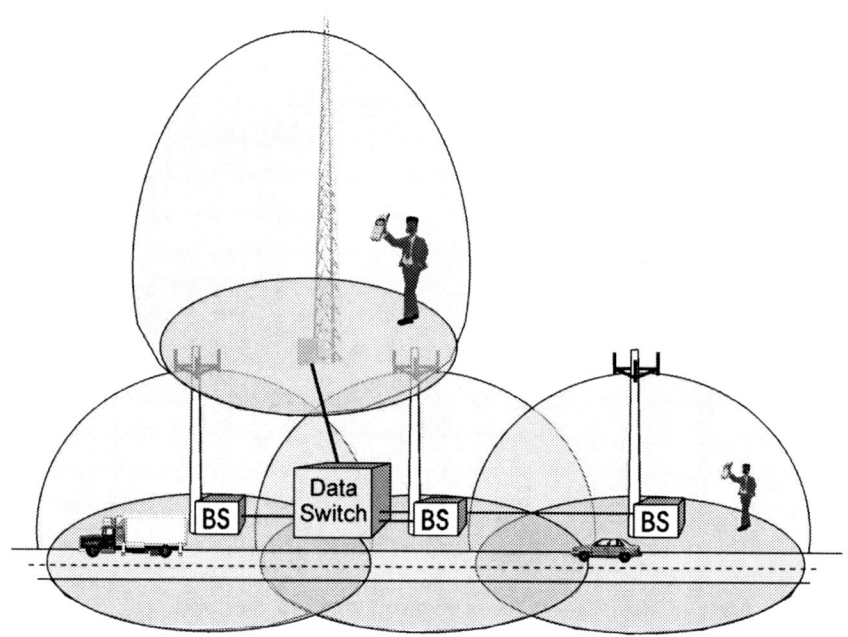

Figure 1.11, WiMax Mobile Communication Services

WiMAX Standards

WiMAX standards are the operational descriptions, procedures or tests that allow manufacturers to produce devices that reliably operate and can work with devices produced by other manufacturers. The development of WiMAX standards is overseen by the Institute of Electrical and Electronics Engineers (IEEE).

The WiMAX standard is given the standard identifier of 802.16 (there are several variations of 802.16). 802.16 specifications primarily cover the lower layers including the physical layer and media access control (MAC) layer and define the different levels of quality of service (QoS) that can be provided.

Even within the WiMAX specification, there are multiple radio interface types. The need for these variations is typically the result of different industry requirements such as fixed point to point communications or mobile broadband communications, which require tradeoffs in radio access types in exchange for key requirements such as mobility, higher speed data communication rates or longer transmission distance.

To enable the 802.16 system to provide mobile operation (not fixed location), the 802.16e specification was created and released in 2006. 802.16e defines mobile broadband operation in the 2-6 GHz frequency range. 802.16e added mobility management, extensible authentication protocol (EAP), handoff (call transfer), and power saving sleep modes to the WiMAX system.

In addition to the basic 802.16 radio interface specifications, several other 802.16 specifications have been created to the standards of the network and operation of WiMAX devices and systems.

802.16f is an IEEE standard that defines the network management information base (MIB) parts that are used for the 802.16 WiMAX system. 802.16g is an IEEE standard that defines the management processes (management plane) that are used for the 802.16 WiMAX system. 802.16.2-2004 is an IEEE standard that defines how to plan and setup broadband radio transmitters in 802.16 systems so they can co-exist.

Figure 1.12 offers a comparison between the original fixed WiMAX standard and the WiMAX standard that can be used for fixed, mobile and portable. This table shows that the original 802.16 standard was released in 2004 and it was only capable of providing fixed wireless data services. It used OFDM modulation and could be deployed in both TDD or FDD formats. The 802.16e standard was released in 2005 (now merged into the original 802.16 standard) was designed for fixed, mobile and portable operation. It used OFDMA modulation with TDD and optionally FDD duplexing capability.

Characteristic	Fixed WiMax	Mobile WiMax
Industry Standard	802.16-2004	802.16e-2005
Access Type	Fixed	Fixed, Portable and Mobile
Modulation	OFDM	OFDMA
Duplexing	TDD, FDD	TDD, FDD Optional
Handoffs	No	Yes
Types of Service Providers	DSL, Cable Modems and Competitive Access Providers (CAPs)	Mobile Operators, DSL, Cable Modems, Wireless and Wired ISPs
Subscriber Units	High Performance Outdoor and Indoor CPE	Low Cost Consumer Electronics CPE and Embedded Modules
Preferred Frequency Bands	2.5 GHz, 3.4-3.6 GHz 5.8 GHz	2.3-2.4 GHz, 2.5-2.7 GHz, 3.3-3.4 GHz, 3.4-3.8 GHz

Figure 1.12, WiMax Standards

Figure 1.13 depicts the evolution of the 802.16 wireless broadband specification over time. The original 802.16 specification offered fixed wireless broadband service at 10-66 GHz. To provide fixed wireless broadband in the 2-11 GHz range, the 802.16a specification was created. Additional variations of the original 802.16 specification were created until in 2004, these specifications were merged into a single 802.16-2004 main specification. Since the 802.16-2004 specification was released, an 802.16e addendum was approved that adds portability to the 802.16 WiMAX system.

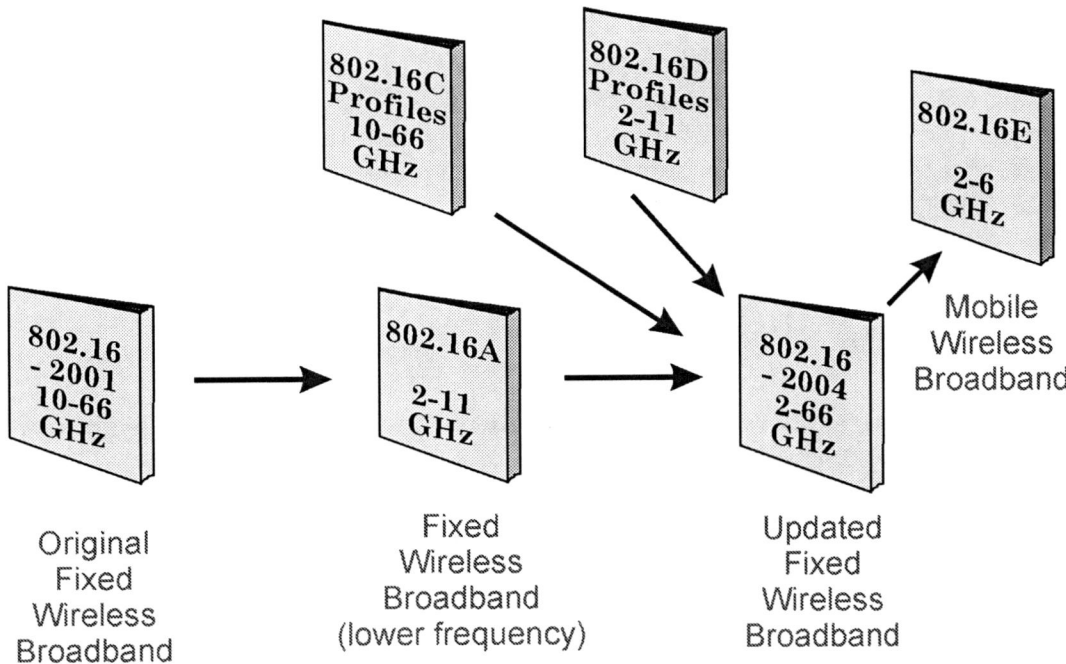

Figure 1.13, Wireless Broadband 802.16 Evolution

WiMAX Broadband Applications

Broadband communication service is the transfer of digital audio (voice), data, and/or video communications at rates greater than wideband communications rates (above 1 Mbps). Broadband connections allow for the providing of multiple services such as telephone (voice), data, and video on one network.

WiMAX VoIP

Digital telephony is a communication system that uses digital data to represent and transfer analog signals. These analog signals can be audio signals (acoustic sounds) or complex modem signals that represent other forms of information.

WiMAX systems can provide telephone services through the use of IP Telephony (voice over Internet protocol – VoIP). These IP networks initiate, process, and receive voice or digital telephone communications using IP protocol. WiMAX systems can provide digital telephone service through the use of analog telephone adapters (ATAs) or IP telephones. ATAs convert IP signals into standard telephone (dial tone) formats.

Broadband Data Connections

Broadband data connections are the transmission digital data signals above 1 Mbps for consumer connections and over 45 Mbps or higher for LANs, MANs and WANs. When WiMAX service providers offer broadband data services that can connect to the Internet, the operator may be called a wireless Internet service provider (WISP).

Digital Television

Digital television is the process or system that transmits video images through digital transmission. Digital transmission is divided into channels (usually compressed) for digital video and audio. Video compression commonly uses one of the motion picture experts group (MPEG) standards to reduce the data transmission rate by a factor of 200:1.

When digital television services are provided by using Internet protocol, it is called IP television (IPTV). IPTV systems initiate, process, and receive television programming using IP protocol. WiMAX systems can provide digital television service through the use of IP set top boxes (IP STBs). IP STBs convert IP signals into standard television formats.

E1/T1 over WiMAX

E1/T1 over WiMAX is the set of processes that are used by service providers to provide E1/T1 services using a WiMAX system. WiMAX systems can replace existing E1/T1 lines (bypass) or they can be used to provide new E1/T1 lines (or added capacity). WiMAX links can be an optimal way to extend the reach of fiber rings to high-usage customers.

Urban WiMAX Hot Zones

Hot zones are geographic regions or service access points that offer connectivity to devices or people who have compatible access devices (e.g. WiMAX card) and are authorized to use the services. Since WiMAX hot zones are relatively large and multiple sites are managed by a single operator or operators that allow roaming visitors, WiMAX hot zone customers consistently get high-speed data services.

Surveillance Services

Surveillance services are the capturing of video and other information for the observation of an area or location at another location. The growing awareness and needs for security, public safety, crime prevention, and asset protection are increasing the demand for new lines for video cameras and other monitoring equipment. WiMAX systems can be used to rapidly deploy video surveillance systems. The ability to communicate common IP protocols allows for the use of proven and inexpensive IP based surveillance devices and systems.

Using licensed WiMAX systems for surveillance services ensures there is no radio interference, ensures good reliability and because it is wireless, this enables the surveillance system to continue to operate in all types of weather and disaster conditions.

Multi-Tenant Units (MTU) and Multi-Dwelling Unit (MDU) Connections

Multi-tenant units and multi-dwelling unit connections provide communication services to buildings that are divided into areas that are occupied by multiple user groups. A single high-speed data connection can be used to provide a mix of voice, data and video services to different groups of users. The sharing of a single high-speed data connection allows for the reduction of equipment and operational cost per user. Building owners may decide to provide these high-speed data services as part of the rental package giving them a competitive advantage over other building owners. WiMAX systems can be an ideal solution for providing high-speed data services to older buildings in locations that are not accessible for cable installation.

Rural Connections

Rural connections are communication links that are available to people or companies that are located in sparsely populated areas. Many governments (including the United States) subsidize the development and deployment of services to users who live in rural areas. Compared to wired systems, WiMAX wireless systems can provide services to rural areas at a much lower cost.

Wireless Broadband System Parts

Wide-area wireless broadband systems are typically composed of end user subscriber stations (access devices), base stations (access nodes), packet switches and gateways.

Network Topology

Network topology is the physical and logical relationship between nodes in a network as well as the layout and structure of a network. The WiMAX system can be setup as a point to point (PTP), point to multipoint (PMP) or a mesh network.

Point to Point (PTP)

Point to point communication is the process of transferring information from one device (or point) to one other device (single receiving point). The WiMAX system can use PTP communication for high-speed communication links for backhaul (system interconnection) applications.

Point to Multipoint (PMP)

Point to multipoint communication is the process of transferring information from one device (or point) to multiple points or devices (multiple receiving points). The WiMAX system can use PMP to provide broadband access to multiple users per base station.

Mesh Network

A mesh network is a communication system where each communication device (typically a computer) is interconnected to multiple nodes (connection points) in the network allowing data packets to travel through alternate paths to reach their destination.

Some or all of the resources of a WiMAX system can be configured to provide mesh network services so that the need to interconnect base stations to access points (such as Internet gateways) can be reduced or eliminated. When a WiMAX system is setup as a mesh network, packets can hop across neighboring base stations to reach other points in the network.

WiMAX transceivers that are part of a mesh network are called nodes. Each mesh node is assigned a unique NodeID and each link between nodes is assigned a LinkID. Packets that enter into the mesh network contain their mesh network destination address (NodeID) and their current mesh link address (LinkID). A mesh node receives and forwards packets towards their destination (NodeID). As nodes in the network transfer packets towards their destination, they change the LinkID within the packet to reflect the next link that will be used in the mesh network.

Mesh nodes that can communicate with each other are part of a neighborhood. Neighborhoods can be small with neighbors that are directly adjacent to each other (immediate neighborhood) or they can be part of a larger neighborhood where nodes must communicate through other nodes so packets can reach their destinations (extended neighborhood).

Mesh node operation may be independent (distributed) or controlled (scheduled) by another network device. When a mesh network is centrally coordinated, one of the nodes is designed as a master synchronization node. The master synchronization node receives requests from mesh nodes, analyzes the bandwidth and transmission requirements and distributes coordinating (scheduling) to other mesh nodes within the network. Mesh nodes that are part of a coordinated mesh system are called a mesh cell.

Mesh nodes regularly broadcast neighbor lists which contain information about available link connections that can be used as part of the mesh network and their associated scheduling times. These lists contain link quality, burst profile, RF power levels and control slot information that will be used to gain access and transmit on the link.

Mesh networks can be setup as either logical or physical connections. A logical mesh network is a system that uses existing WiMAX transceivers to receive and forward packets or data towards its destination. For logical mesh networks, links can be dynamically setup and removed as desired by the mesh network. Physical mesh nodes use transceivers that have their links pre-established so they are not addressable (cannot have their addresses dynamically assigned).

When a WiMAX device wants to attach to a mesh network (gain network entry), it must initially search to see if radio channels are available for a mesh network. If it finds an available radio channel, it needs to communicate with a mesh node that is willing to help it attach to the mesh network (called a sponsor node). The sponsor node will allow it to make a radio connection and negotiate basic communication parameters. It will then relay the request to join the mesh network to the node or part of the network that will authorize and assign resources to the new mesh node (e.g. the master synchronization node).

Figure 1.14 shows a WiMAX system configured as a mesh network. The base station (BS) devices in a WiMAX system are part of a neighborhood domain and the devices within this neighborhood can be configured as relay point nodes so they can help transfer packets from their neighbors towards their destination. A distant subscriber station that cannot directly communicate with the mesh base station can send packets to a neighbor that can relay the packets to the mesh base station that connects to Internet gateway. This

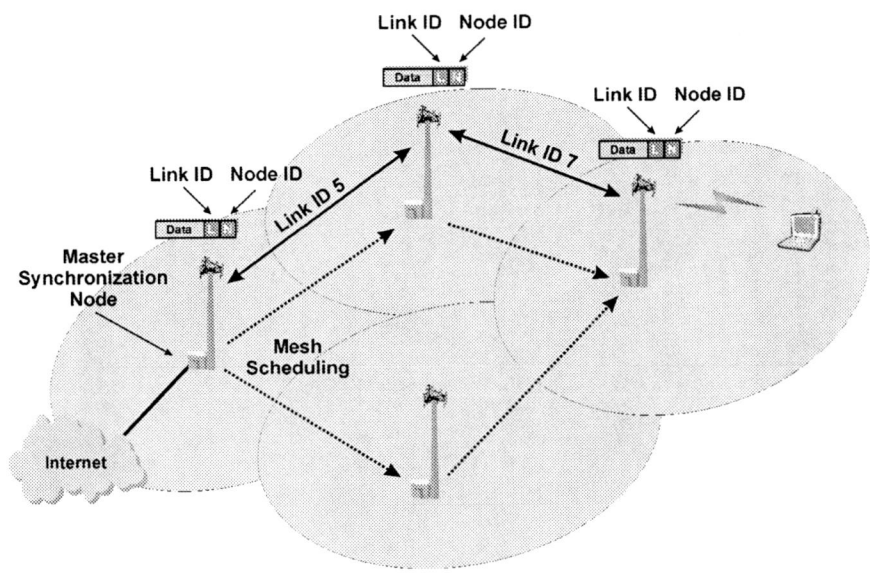

Figure 1.14, WiMax Mesh Network

example shows that a mesh network contains links between mesh nodes that are uniquely identified by link identifiers (LinkID) and that the destination point for packets traveling through the mesh network are identified by a node identifier (NodeID). Packets that travel through the mesh network contain both the destination address (NodeID) and the current link address (LinkID). As packets travel through mesh nodes, the link address changes.

A directed mesh network is communication system where each communication device (typically computers) is interconnected to multiple nodes (connection points) through the use of highly directional connections (e.g. directional antennas). In directed mesh networks, data packets travel through the directed paths to reach their destination. WiMAX systems may use a directed mesh network to interconnect areas to a central facility or gateway (such as a gateway to the Internet).

Figure 1.15 depicts a WiMAX system that is setup as a directed mesh network. This diagram shows how a WiMAX operator can use directional antennas that interconnect to base stations to allow packets to be relayed from distant locations to a central location (such as an Internet gateway) using

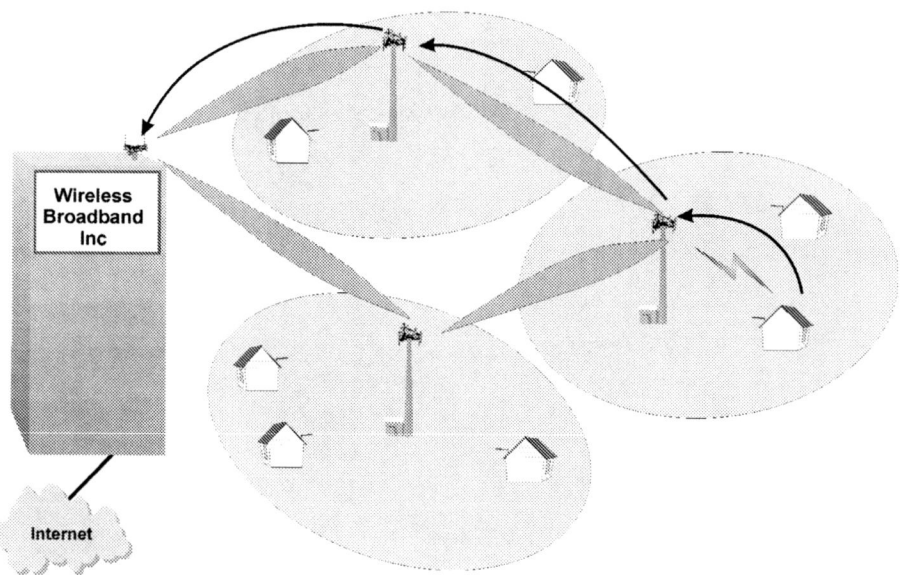

Figure 1.15, WiMax Directed Mesh Network

WiMAX frequencies. This example shows that the use of directional antennas results in highly focused transmission beams which have little impact on the overall radio coverage of the WiMAX system.

Chassis Based Systems

Chassis based systems allow for the insertion or attachment of modules or assemblies to add functionality or capacity to systems or networks.

Pico Based Systems

Pico based systems are self-contained assemblies that contain all the needed functionality of a system or portion of a network.

Subscriber Stations (SS)

Subscriber stations (SS) in a WiMAX system are transceivers (transmitter and receivers) that convert radio signals into digital signals that can be routed to and from communication devices. The types of WiMAX subscriber stations range from portable PCMCIA cards to fixed stations that provide service to multiple users.

Subscriber stations may be capable of full frequency division duplex (FDD), half frequency duplex (H-FDD), time division duplex (TDD) or any combination of these access types.

Figure 1.16 shows the different types of WiMAX access devices. WiMAX access devices include external boxes that connect to an Ethernet or USB port, PCMCIA card with external antennas or portable devices with built in radio modems.

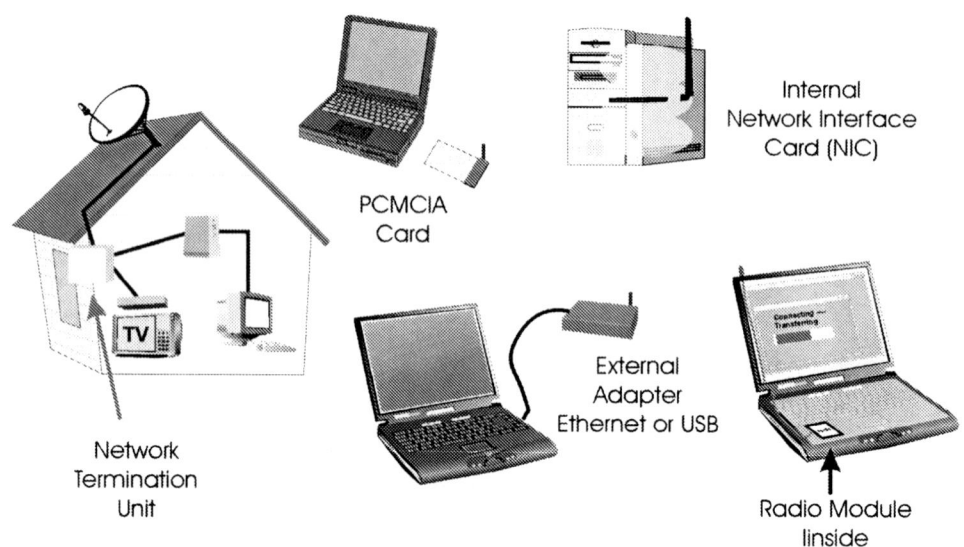

Figure 1.16, WiMax Devices

Indoor Subscriber Stations

Indoor subscriber stations are devices or assemblies that enable users to receive and convert radio or communication signals that are enclosed and/or have connections that allow them to be used in indoor environments.

Outdoor Subscriber Stations

Outdoor subscriber stations are devices or assemblies that enable users or devices to receive and convert radio or communication signals that are enclosed and/or have connections that allow them to be used in outdoor environments.

Base Stations (BS)

A base station is a radio access transceiver (combined transmitter and receiver) that is used to connect subscriber stations to a WiMAX system. Base stations convert and control the sending of data packets and can connect one or many wireless devices to a backbone network.

Base stations can perform one or many types of data transfer functions including bridging (linking networks), retransmitting (repeating), distributing (hubs), directing packets (switching or routing) or to adapt formats for other types of networks (gateways).

Indoor Base Stations

Indoor base stations are devices or assemblies that enable systems or system connections to receive and convert radio or communication signals that are enclosed and/or have connections that allow them to be used in indoor environments.

Outdoor Base Stations

Outdoor base stations are devices or assemblies that enable systems or system connections to receive and convert radio or communication signals that are enclosed and/or have connections that allow them to be used in outdoor environments.

Packet Switches

A packet switch is a device in a data transmission network that receives and forwards packets of data. The packet switch receives the packet of data, reads its address, searches in its database for its forwarding address, and sends the packet toward its next destination.

Packet switching is different than circuit switching because circuit switching makes continuous path connections based on a signal's time of arrival (TDM) port of arrival (cross-connect) or frequency of arrival. In a packet switch, each transmission is packetized and individually addressed, much like a letter in the mail. At each post office along the way to the destination, the address is inspected and the letter forwarded to the next closest post office facility. A packet switch works much the same way.

Operational Support System (OSS)

Operations support systems are combinations of equipment and software that are used to allow a network operator to perform the administrative portions of the business. These functions include customer care, inventory management and billing. Originally, OSS referred to the systems that only supported the operation of the network. Recent definition includes all systems required to support the communications company including network systems, billing, and customer care.

Gateways

Gateways are communication devices that transform data received from one network into a format that can be used by a different network. A gateway usually has more intelligence (processing function) than a bridge as it can adjust the protocols and timing between two dissimilar computer systems or data networks. A gateway can also be a router when its key function is to switch data between network points.

Wireless gateways are access points that can assign temporary IP addresses (DHCP) and have the ability to share a single public IP addresses with several private IP addresses (NAT).

Antennas

An antenna is a device used to convert signals between electrical and electromagnetic form. Antennas are usually designed to operate over a specific frequency range. Directional antennas are designed to focus (concentrate) the transmitted energy in a particular direction to achieve antenna gain.

Technologies

Some of the key technologies used in WiMAX systems include orthogonal frequency division multiplexing, frequency reuse, adaptive modulation, diversity transmission and adaptive antennas.

Orthogonal Frequency Division Multiplexing (OFDM)

OFDM is a process of transmitting several high speed communication channels through a single communication channel using separate sub-carriers (frequencies) for each radio channel. The use of OFDM reduces the effects of multi-path and delay spread, which is especially important for lower frequencies and near line of sight (NLOS) transmission.

Multi-path propagation is the transmission of a radio signal which travels over two or more paths from a transmitter to a receiver. Multi-path transmission can cause changes in the received signal level as delayed signals can either add or subtract from the received signal level. Multi-path is not usually a challenge on systems that use higher frequencies as these systems tend to use highly directional (high-gain) antennas for direct line of sight transmission.

Multi-path propagation is frequency dependent meaning that the multiple paths radio signals travel will vary depending on its' frequency.

Figure 1.17 illustrates how a transmitted signal may travel through multiple paths before reaching its destination. In this example, the same signal is reflected off an office building where it is received by the subscriber device. The reflected signal is delayed (travels a longer path) and subtracts from the direct signal resulting in a dead spot (fade) at the receiver. Furthermore, mutli-path propagation is sensitive to frequency and that distortion occurs at different points when other frequencies are used. When a different frequency is used, the reflected signal is redirected and it does not subtract from the direct signal.

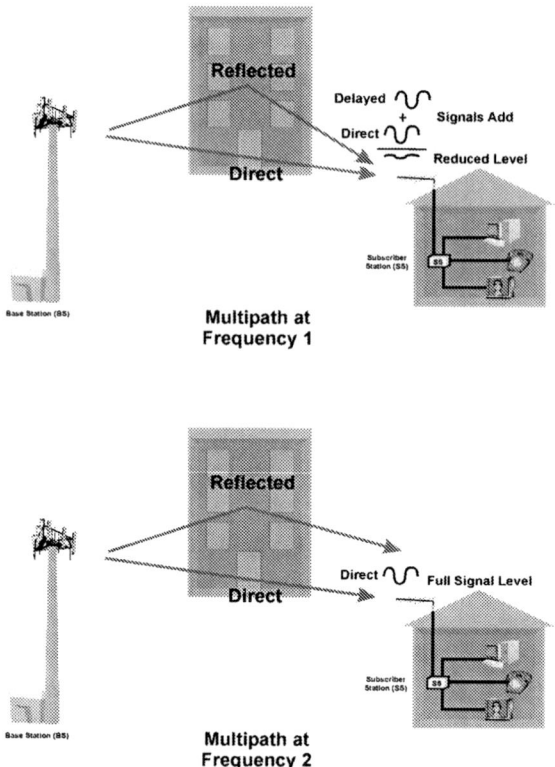

Figure 1.17, Multi-path Propagation

For a wide radio channel that is divided into several sub-carriers, each sub-carrier channel operates at a different frequency and can have different transmission characteristics than other sub-carriers. Because multiple sub-carriers are typically combined for a single subscriber, this can reduce the effects of multi-path fading.

The use of multiple sub-carriers also has the effect of reducing the symbol rate, which can reduce the effects of delay spread. Delay spread is a product of multi-path propagation where symbols become distorted and eventually overlap due to the same signal being received at a different time. It becomes a significant problem in mountainous areas where signals are reflected at great distances. Delay spread can be minimized by either using an equalizer to adjust for the multi-path distortions or to divide a communication channel into sub-carriers (e.g. OFDM) where each sub-carrier transfers data at a much slower data transmission rate thereby reducing the effects of delay spread.

Figure 1.18 demonstrates how OFDM divides a single radio channel into multiple coded sub-channels. A high-speed digital signal is divided into multiple lower-speed sub channels that are independently from each other and can be individually controlled. The OFDM process allows bits to be sent on multiple sub channels. The channels selected can be varied based on the quality of the sub channel. In this figure, a portion of a sub channel is lost due to a frequency fade. As a result of the OFDM encoding process, the missing bits from one channel can be transmitted on other channels.

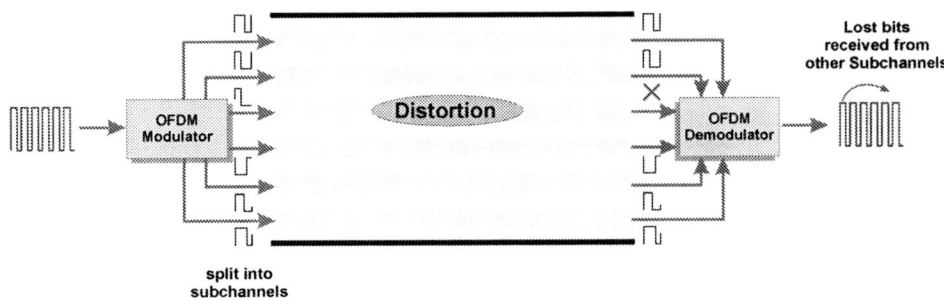

Figure 1.18, Orthogonal Frequency Division Multiplexing (OFDM)

Orthogonal Frequency Division Multiple Access (OFDMA)

Orthogonal frequency division multiple access is the process of dividing a radio carrier channel into several independent sub-carrier channels that are shared between simultaneous users of the radio carrier. When a mobile radio communicates with an OFDMA system, it is dynamically assigned a specific sub-carrier channel or group of sub-carrier channels within the radio carrier. By allowing several users to use different sub-carrier channels, OFDMA systems increase their ability to serve multiple users and the OFDMA system may dynamically allocate varying amounts of transmission bandwidth based on how many sub-carrier channels have been assigned to each user.

As demonstrated in figure 1.19, the WiMAX system allows more than one simultaneous user per radio channel through the use of orthogonal frequency division multiple access (OFDMA). The WiMAX radio channel can be divided into multiple sub-carriers and that the sub-carriers can be dynamically assigned to multiple users who are sharing a radio carrier signal. Finally, the data rates that are provided to each user can vary based on the number of sub-carriers that are assigned to each user.

Figure 1.19, Orthogonal Frequency Division Multiple Access (OFDMA)

Frequency Reuse

Frequency reuse is the process of using the same radio frequencies on radio transmitter sites within a geographic area that are separated by sufficient distance to cause minimal interference with each other. Frequency reuse allows for a dramatic increase in the number of customers that can be served (capacity) within a geographic area on a limited amount of radio spectrum (limited number of radio channels). Frequency reuse allows WiMAX system operators to reuse the same frequency at different cell sites within their system operating area.

The number of times a frequency can be reused is determined by the amount of interference a radio channel can tolerate from nearby transmitters that are operating on the same frequency (carrier to interference ratio).

Carrier to interference (C/I) level is the amount of interference level from all unwanted interfering signals in comparison to the desired carrier signal. The C/I ratio is commonly expressed in dB. Different types of systems can tolerate different levels of interference dependent on the modulation type and error protection systems. The typical C/I ratio for narrowband mobile radio systems ranges from 9 dB (GSM) to 20 dB (analog cellular). WiMAX systems can be much more tolerant to interference levels (possibly less than 3 dB C/I) when OFDM and adaptive antenna systems are used.

WiMAX systems may also reuse frequencies through the use of cell sectoring. Sectoring is a process of dividing a geographic region (such as a radio coverage area) where the initial geographic area (e.g. cell site coverage area) is divided into smaller coverage areas (sectors) by using focusing equipment (e.g. directional antennas).

Figure 1.20 shows how radio channels (frequencies) in a WiMAX communication system can be reused in towers that have enough distance between them.

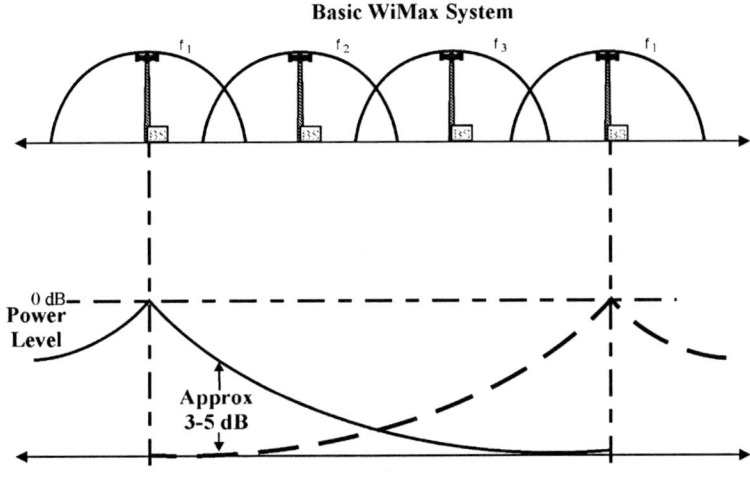

Figure 1.20, WiMax Frequency Reuse

The radio channel signal strength decreases exponentially with distance. As a result, mobile radios that are far enough apart can use the same radio channel frequency with minimal interference.

Modulation

Modulation is the process of changing the amplitude, frequency, or phase of a radio frequency carrier signal with the information signal (such as voice or data). The 802.16 system uses different types of digital modulation depending on a variety of transmission factors. The modulation types used in 802.16 systems include binary phase shift keying (BPSK), quadrature phase shift keying (QPSK) and quadrature amplitude modulation (QAM).

Binary Phase Shift Keying (BPSK)

Binary phase shift keying (BPSK) is a modulation process that converts binary bits into phase shifts of the radio carrier without substantially changing the frequency of the carrier waveform. The phase of a carrier is the relative time of the peaks and valleys of the sine wave relative to the time of an unmodulated "clock" sine wave of the same frequency. BPSK uses only two-phase angles, corresponding to a phase shift of zero or a half cycle (zero or 180 degrees of angle). WiMAX uses BPSK modulation when a very robust signal is required.

Quadrature Phase Shift Keying (QPSK)

Quadrature phase shift keying (QPSK) is a type of modulation that uses 4 different phase shifts of a radio carrier signal to represent the digital information signal. These shifts are typically +/- 45 and +/- 135 degrees.

Quadrature Amplitude Modulation (QAM)

QAM is a combination of amplitude modulation (changing the amplitude or voltage of a sine wave to convey information) together with phase modulation. There are several ways to build a QAM modulator. In one process, two modulating signals are derived by special pre-processing from the information bit stream. Two replicas of the carrier frequency sine wave are generated; one is a direct replica and the other is delayed by a quarter of a cycle (90 degrees). Each of the two different derived modulating signals are then used to amplitude modulate one of the two replica carrier sinewaves respectively. The resultant two modulated signals can be added together. The result is a sine wave having a constant unchanging frequency while having an amplitude and phase that both vary to convey the information. At the detector or decoder the original information bit stream can be reconstructed. QAM conveys a higher information bit rate (bits per second) than a BPSK or QPSK signal of the same bandwidth, but is also more affected by interference and noise.

Figure 1.21 shows that amplitude and phase modulation (QAM) can be combined to form an efficient modulation system. One digital signal changes the phase and another digital signal changes the amplitude.

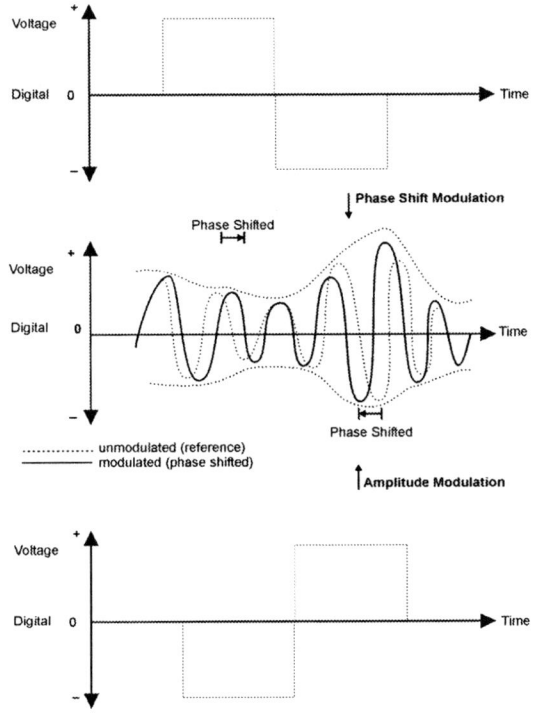

Figure 1.21, Quadrature Amplitude Modulation (QAM)

Adaptive Modulation

Adaptive modulation is the process of dynamically adjusting the modulation type of a communication channel based on specific criteria (e.g. interference or data transmission rate). WiMAX systems use adaptive modulation to ensure the modulation type matches the channel characteristics (signal quality level).

In general, the more efficient (data transmission capacity) the modulation type, the more complex or precise the modulation process is. The more precise the modulation process (smaller changes represent digital bits), the more sensitive the modulation is to distortion or interference. This usually means that as the data transmission rate increases, the sensitivity to interference intensifies. To help manage this process and ensure the maximum data transmission rate possible, 802.16 systems automatically change their data modulation types and data transmission rates (Autorate) based on the ability of the channel to transfer data. The 802.16 systems will usually try to send information at the highest data transmission rate possible. If the data transmission rate cannot be maintained, the 802.16 systems will attempt to transmit at the next lower data transmission rate. Lower data transmission rates generally use a less complex (more robust) modulation type.

Diversity Transmission

Diversity transmission is the process of using two or more signals to carry the same information source between a transmitter and a receiver. Diversity transmission can use the physical separation of antenna elements (spatial diversity), the use of multiple wavelengths (frequency diversity) and the shifting of time (time diversity).

Protocols on the WiMAX system are designed to take advantage of diversity transmission options and to allow for the use of multiple input multiple output (MIMO) antenna systems. MIMO is the combining or use of two or more radio or telecom transport channels for a communication channel through the user of multiple antenna elements. The use of MIMO to combine alternate transport links provides for higher data transmission rates (inverse multiplexing) and increased reliability (interference control).

Transmission Diversity

Transmission diversity is the process of sending two or more signals from the same information source so a receiver can select or combine the signals to produce a received signal of better quality than a single transmitted signal.

Receive Diversity

Receive diversity is used to select or combine a received signal to yield a stronger signal quality level. Receive diversity uses two antennas that are physically separated vertically or horizontally.

Receiver diversity can compensate for radio signal fading that may occur on a single antenna, and may be performed by maximum ratio combining (MRC) or selection diversity. Maximal ratio combining is the process of combining the signals from two or more antenna elements to increase the level and quality of a received signal. Selection diversity is the process of selecting one antenna from a set of receiving antennas to increase the level or quality of a received signal.

Frequency Diversity

Frequency diversity is the process of receiving a radio signal or components of a radio signal on multiple channels (different frequencies) or over a wide radio channel (wide frequency band) to reduce the effects of radio signal distortions (such as signal fading) that occur on one frequency component but do not occur (or are not as severe) on another frequency component.

Temporal (Time) Diversity

Time diversity is the process of sending the same signal or components of a signal through a communication channel where the same signal is transmitted or received at different times. The reception of two or more of the same signal with time diversity may be used to compare, recover, or add to the overall quality of the received signal.

Spatial Diversity

Spatial diversity is a method of transmission or reception employed to minimize the effects of fading by the simultaneous use of two or more antennas spaced a number of wavelengths apart.

Antenna diversity is a form of spatial diversity that improves the reception of a radio signal by using the signals from two (or more) antennas to minimize the effects of radio signal fading or distortion. Antenna diversity typically requires the antennas to be spaced a number of wavelengths apart.

Space time coding is the adding of time information to transmission carriers to allow diversity operation by identifying and processing multiple carriers of the same signal that may arrive at different times and or from different locations.

Figure 1.22 shows different types of diversity transmission and reception. The antenna (spatial) diversity utilizes the distance between antennas to improve signal performance. Frequency (spectral) diversity transmits the same or related information on multiple frequency signals to reduce frequency selective fading. Time (temporal) diversity overcomes the challenges of burst distortion by allowing the same information signal to be received at different times.

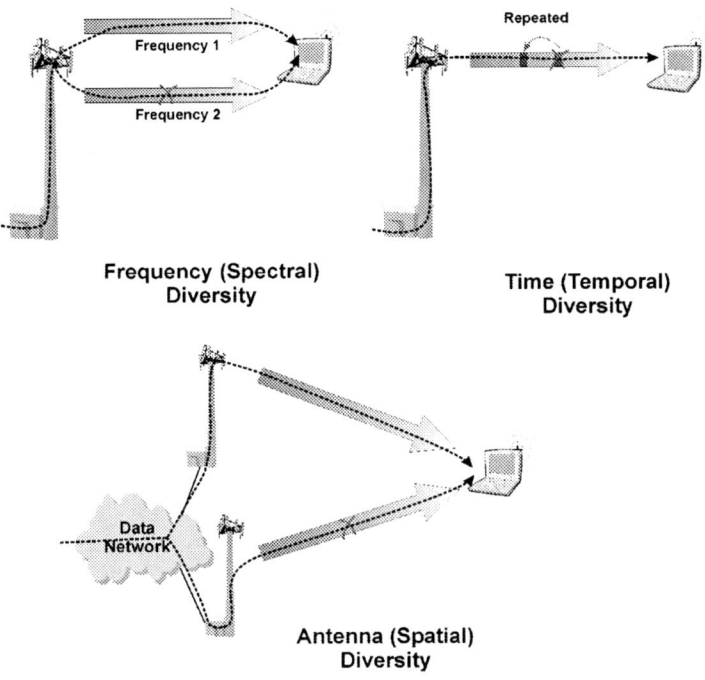

Figure 1.22, Diversity Transmission

Adaptive Antenna System (AAS)

An adaptive antenna system allows a transmitter to focus radio beams to increase the transmission range, reduce interference and increase signal quality. When an AAS system is used to allow multiple users to communicate with the same transceiver (multiple beams), it is called spatial division multiple access (SDMA). SDMA technology has been successfully used in satellite communications for several years. In some SDMA systems, radio beams may dynamically change with the location of the mobile radio.

The WiMAX system is designed with AAS capability. To support AAS, it is necessary to supplement the medium access control (MAC) protocol with additional commands so that base stations can better monitor subscriber stations which may be operating in a narrow focused beam area. If the subscriber station were to move out of the focused beam area, the system could loose control of the subscriber station.

Figure 1.23 shows an example of a WiMAX adaptive antenna system (AAS). The cell site can focus radio signals using the same frequency to multiple devices within the same cell site. Focusing of the radio signal allows for an increase in the distance that a cell site can have when communicating with devices. Using AAS technology, the system can adapt the direction of the focused beam to a specific device as it moves throughout the coverage area.

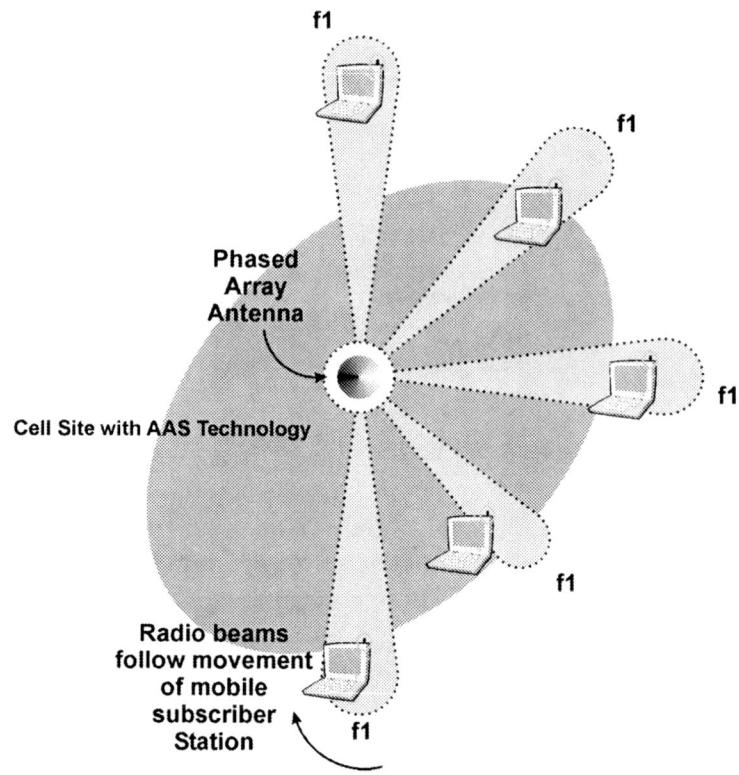

Figure 1.23, WiMax Adaptive Antenna System

WiMAX Radio Overview

A WiMAX radio channel is a communications channel that uses radio waves to transfer information from a source to a destination. It may transport one or many communication channels and communication circuits on a single RF channel.

WiMAX radio channels may operate within different frequency bands, have different radio channel bandwidths, dynamically change modulation types, use a variety of access technologies and other characteristics that allow WiMAX to reliably provide a variety of types of communication services.

The WiMAX radio systems can use a single carrier (SC) or multi-carrier (MC) transmission. Single carrier transmission is the use of a single carrier wave that is modified to carry (transport) all of the information. Multi-carrier is a communication system that combines or binds together two or more communication carrier signals (carrier channels) to produce a single communication channel. This single communication channel has capabilities (capacity) beyond any of the individual carriers that have been combined. When each of the carriers in a multi-carrier system is mutually independent (orthogonal) to each other, it is called orthogonal frequency division multiplexing (OFDM).

Figure 1.24 shows the key components of a basic WiMAX radio system. The major component of a WiMAX system include subscriber station (SS), a base station (BS) and interconnection gateways to datacom (e.g. Internet) and telecom (e.g. PSTN). An antenna and receiver (subscriber station) in the home or business converts the microwave radio signals into broadband data signals for distribution. In the example, a WiMAX system is being used to provide telephone and broadband data communication services. When used for telephone services, the WiMAX system converts broadcast signals to an audio format (such as VoIP) for distribution to IP telephones or analog telephone adapter (ATA) boxes. When WiMAX is used for broadband data, the WiMAX system also connects the Internet through a gateway to the Internet. The WiMAX system can reach distances of up to 50 km when operating at lower frequencies (2-11 GHz).

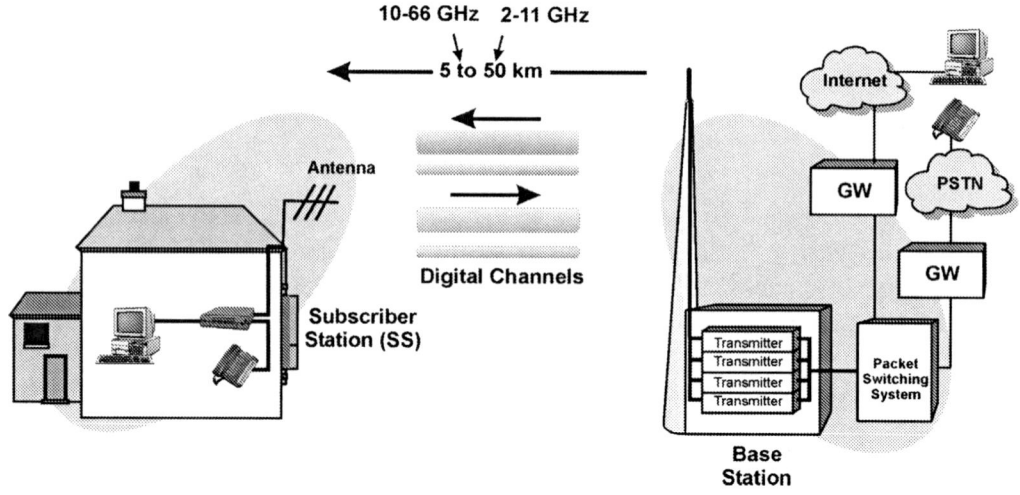

Figure 1.24,WiMax System

WiMAX Protocol Layers

Protocol layers are a hierarchical model of network or communication functions. The divisions of the hierarchy are referred to as layers or levels, with each layer performing a specific task. In addition, each protocol layer obtains services from the protocol layer below it and performs services to the protocol layer above it. The WiMAX system has four key layers; MAC convergence, MAC layer, MAC privacy and the physical layer.

MAC Convergence

The MAC convergence layer is a functional process within a communication device or system that adapts one or more transmission mediums (such as radio packet or circuit data transmission) to one or more alternative transmission formats (such as ATM or IP data transmission).

MAC Layer

The MAC layer is composed of one or more logical communication channels that are used to coordinate the access of communication devices to a shared communications medium or channel (microwave radio). MAC channels typically communicate the availability and access priority schedules for devices that may want to gain access to a communication system.

MAC Privacy

The MAC privacy layer is associated with authenticating and encrypting information over the communication link (inner coding).

Physical Layer

The physical layer performs the conversion of data to a physical transmission medium (such as copper, radio, or optical) and coordinates the transmission and reception of these physical signals. The physical layer receives data for transmission from an upper layer and converts it into physical format suitable for transmission through a network (such as frames and bursts). An upper layer provides the physical layer with the necessary data and control (e.g. maximum packet size) to allow conversion to a format suitable for transmission on a specific network type and transmission line.

Security Sub-Layer

A security sub-layer is the functional process within a communication device or system that performs access controls, identity validation and/or encryption of data.

Figure 1.25 depicts the multiple layers of WiMAX. The physical layer is responsible for converting bits of information into radio bursts while the MAC security layer is responsible for identifying the users (authentication) and keeping the information private (encrypting). The MAC layer is responsible for requesting access and coordinating the flow of information. The MAC convergence layer is used to adapt the WiMAX system to other systems such as ATM, Ethernet or IP data systems.

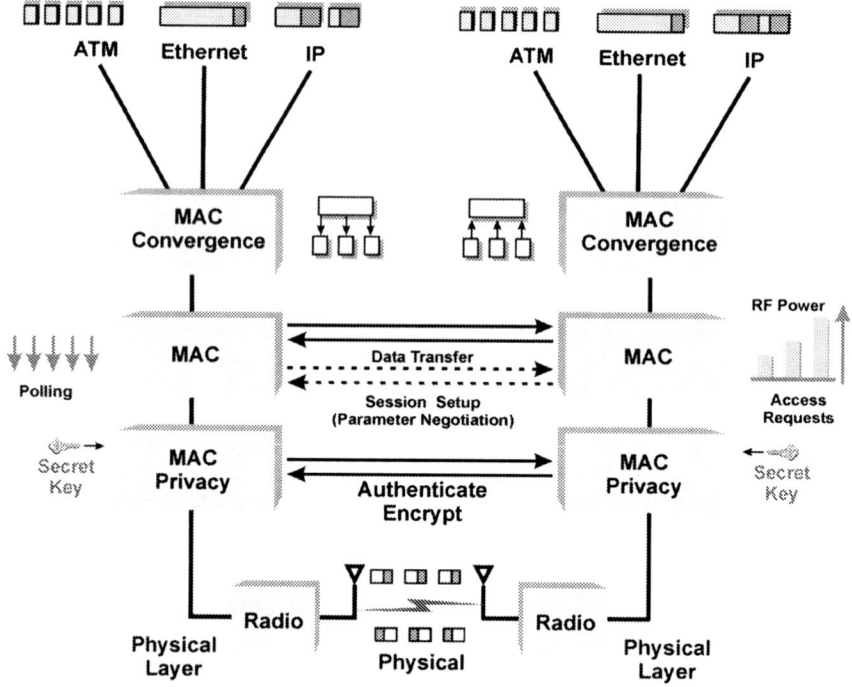

Figure 1.25, WiMax Protocol Layers

Radio Propagation

Radio propagation is the process of transferring a radio signal (electromagnetic signal) from one point to another point. Radio propagation may involve a direct wave (space wave) or a wave that travels along the surface (a surface wave). Radio propagation characteristics typically vary based on the medium of transmission (e.g. Air) and the frequency of radio transmission. WiMAX systems may be operated as line of sight (direct transmission path) or non line of sight (indirect transmission path) systems.

WiMAX systems are typically designed with a radio link budget. A link budget is the maximum amount of signal losses that may occur between a transmitter and receiver to achieve an adequate signal quality level. The link budget typically includes cable losses, antenna conversion efficiency, propagation path loss, and fade margin.

Due to transmission impairments, a fade margin is budgeted in a communication link. Fade margin is the amount of signal loss, usually expressed in decibels, that a radio signal in a communication path is anticipated to change (or budgeted to change). This helps to ensure that typical signal fading periods do not result in a lower than expected quality of service.

One of the key types of signal fades that occur on microwave systems is a rain fade. A rain fade is the signal loss that results from signal absorption and scattering in water droplets (rain).

Line of Sight (LOS)

Line of sight (LOS) is a direct path in a wireless communication system that does not have any significant obstructions. WiMAX systems that operate in the 10-66 GHz range are LOS systems.

Non Line of Sight (NLOS)

Non line of sight (NLOS) is a wireless communication system that does not have a direct path (can have significant obstructions) between the transmitter and receiver. NLOS systems can use radio signals for transmission.

Figure 1.26 shows how non line of sight (NLOS) radio propagation can allow a radio signal to reach its destination in congested areas. A radio tower is transmitting through an urban area which does not allow a radio signal to travel a direct path from the tower to the receiver. Accordingly, multiple alternate paths are reflected off a building to reach its destination. A main signal (shortest reflected signal) and another signal (delayed signal) become part of the received signal.

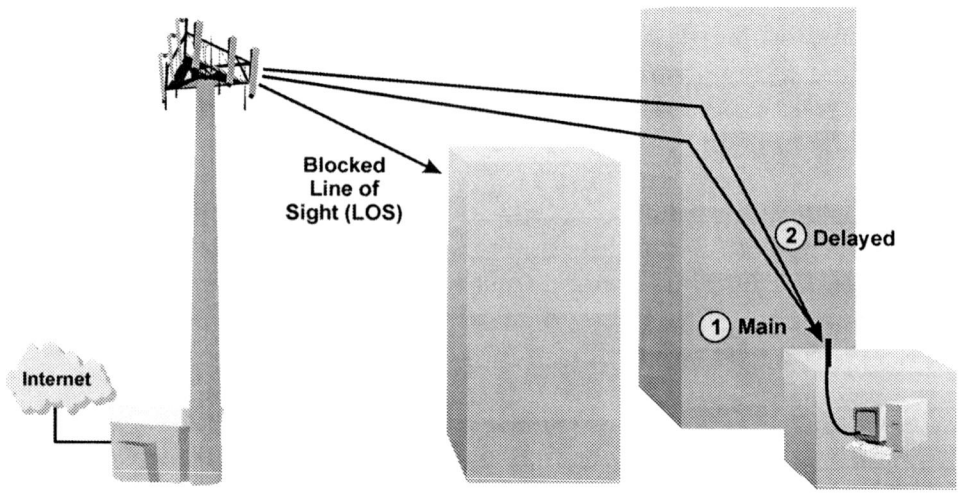

Figure 1.26, Near Line of Sight Radio Propagation

Addressing

Each WiMAX radio is configured at the factory with a unique 48 bit medium access control address (MAC address) as identified in IEEE standard 802-2001. The first few bits of the MAC address indicate the manufacturer of the device. The remaining bits are a unique serial number of the device. While it is possible for a single subscriber device to have more than one 48 bit physical MAC address, most devices have a single MAC address.

The 48 bit MAC address is not part of the transmitted packet (MPDU). Instead, the device is identified by a 16 bit connection identifier (CID) that is assigned after the device has connected to the system. The MAC address is transferred during the device registration or authentication process to allow the system to identify the specific user. A single WiMAX device may contain 1 or more globally unique MAC addresses.

Each WiMAX device typically has several CIDs assigned. Some of the CIDs are used for controlling information and some are used to identify user data transmission channels (traffic channels).

The 16 bit CID is used to identify and categorize traffic (a maximum of 65,535 CIDs can exist per RF channel). Some CIDs are pre-assigned for specific functions (such as initial ranging) and others are unique to a specific connection.

A single CID may be shared by several services (logical channels). Each of these channels is a service flow and is identified by a service flow identifier (SFID).

To reduce the amount of overhead on a WiMAX radio channel (bits dedicated for control purposes), a shortened version of a CID (called a reduced connection identifier) may be used. A reduced connection identifier (RCID) can be 11, 7 or 3 bits long.

Medium Access Control Protocol Data Units (MAC PDUs)

Medium access control protocol data units are a package of data (group of data bits) that contain header, connection address and data protocol information that is used to control and transfer information across a type of medium (such as a radio channel). The WiMAX system MAC PDUs contain a header, which holds the connection identifier along with control information. MAC PDUs may also have payload of data and error checking bits (CRC) bits after the header (e.g. user data). A MAC PDU header contains a header type, encryption control field, payload type and error checking (CRC) code.

A header type is a data field within the packet header that indicates the type of the header. The header type typically indicates the field format of the header and/or sub-headers that are part of the data packet.

Encryption control is the transferring parameters or sending of signaling messages in a data field within the header of a data packet that indicates encryption is used and possibly the type of encryption that is used. An encryption control field in a header typically indicates the payload of the data packet is encrypted.

Payload type is a data field within a packet header that indicates the format of the payload of data including any sub-headers

A cyclic redundancy check indicator is a data field within a packet header that indicates if and how a CRC error check code is used for the packet of data.

Encryption key sequence is an index value that is used to identify the location of a data packet within a sequence of packets to enable the decryption of the packet.

A length field is a data field within a packet header that holds a number or value that indicates the length of a data packet or a block of data.

A connection identifier is a unique name or number that is used to identify a specific logical connection path in a communication system. For the WiMAX system, the connection identifier is a 16 bit code.

A header check sequence is a calculated code that is used to check if the bits within a header have been received correctly during transmission.

Figure 1.27 shows the typical construction of generic WiMAX medium access control packet. This diagram shows that generic WiMAX MAC data packets contain addressing and controlling information and the payload can have variable length. This diagram shows that the MPDU header is 6 bytes long and that the header type indicator is used to determine which, if any, sub-headers may be used.

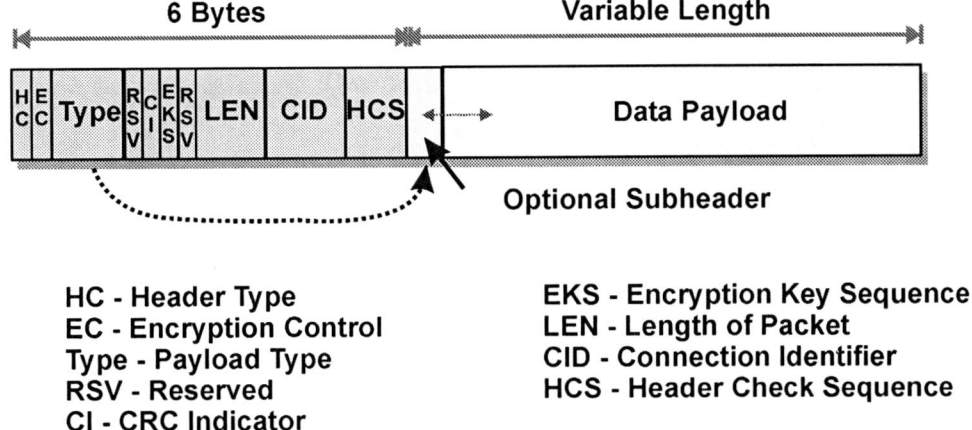

HC - Header Type
EC - Encryption Control
Type - Payload Type
RSV - Reserved
CI - CRC Indicator

EKS - Encryption Key Sequence
LEN - Length of Packet
CID - Connection Identifier
HCS - Header Check Sequence

Figure 1.27, WiMax MAC PDU

Radio Packets (Bursts)

A radio packet is a short transmission (a burst) of information that occurs on a radio channel. Radio bursts contain reference sequences (preamble and possible a midamble), control information and payload of data.

The radio packet burst may have different types of radio characteristics such as modulation type, error coding, preamble length and transmission guard time periods. The combination of these characteristics is called the burst profile.

A burst set is a single transmission (RF packet) that contains a preamble along with one or more bursts of information. The bursts of information contained within the RF packet may have different modulation and coding types. A burst frame is the complete set of information that is contained in a transmission burst.

The bursts within a burst set are sequenced according to their modulation complexity. Bursts with lower complexity modulation types are located at the beginning of the radio packet. Bursts that follow can use modulation types with higher complexity (e.g. QPSK, QAM). This allows subscriber stations to receive and decode all the bursts up to the burst with the highest modulation type it can receive.

RF bursts start with a sequence of bits (a preamble) that the receiving device can recognize and lock onto. Once the receiving device locks onto the preamble, it knows where to find the rest of the packets.

For longer RF bursts, midamble sequences may be periodically inserted to assist receivers in the decoding of bursts. A midamble is a sequence of bits that the receiving device can recognize and lock onto to help decode the bits surrounding the midamble. Increased mobility (speed) can be tolerated when preambles and midambles are sent more frequently.

Burst profiles may continually change on a WiMAX system. Subscriber stations can send burst profile change request messages that define a desired new burst profile characteristic (such as modulation type or error coding). The request may be a result of an increase in the error rate using an existing burst profile.

Data packets may be inserted (embedded) within the payload of a single RF burst, they may be divided (fragmented) so they can be distributed over several radio packet bursts or multiple small packets may be combined (packed) into the payload.

Figure 1.28 depicts an example of a WiMAX radio packet which is made of a preamble and a set of bursts. The figure shows that the modulation type of the burst starts simple (BPSK) and gets more complex as additional bursts follow.

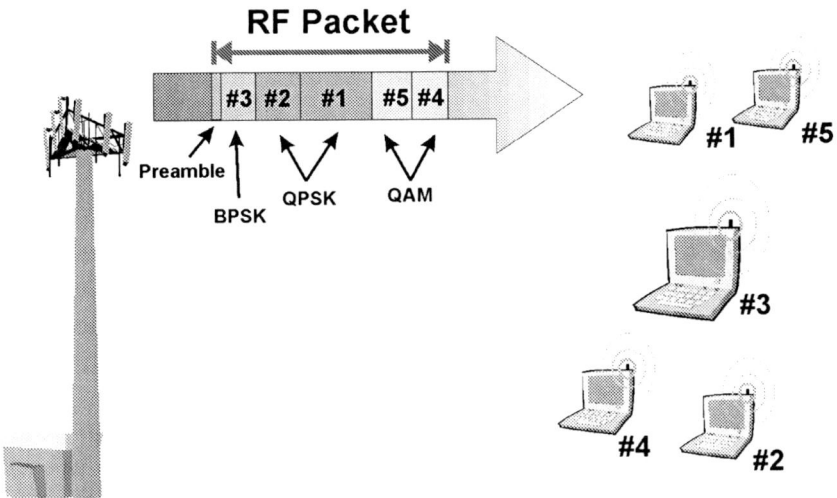

Figure 1.28, WiMax RF Packets

Channel Descriptors

Channel descriptor is a set of information parameters that describe the characteristics associated with a communication channel. The use of a channel descriptor can permit more accurate and successful reception and decoding of information that is sent on a communication channel.

The WiMAX system periodically sends (broadcasts) channel descriptors on the downlink channel to allow the subscriber stations to understand how to decode and transmit messages. Channel descriptors provide information about the uplink and downlink channels.

The downlink channel descriptor contains a downlink frame prefix that provides the information to the receiver about the frame structure of the downlink channel and a downlink map (DL-MAP) that defines what information will be transmitted. The DL-MAP contains a downlink interval usage code (DIUC) that defines when information will be transmitted on the downlink and what formats it is supposed to use (burst profile).

The uplink channel descriptors contain an uplink map (UL-MAP) message that defines when a subscriber station is allowed to transmit on the uplink and what formats it is supposed to use (burst profile). The UL-MAP contains an uplink interval usage code (UIUC) that defines when a subscriber station is allowed to transmit on the uplink and what formats it is supposed to use (burst profile).

Figure 1.29 demonstrates how the WiMAX system uses channel descriptors to define the allocated times and burst types that are used on WiMAX radio channels.

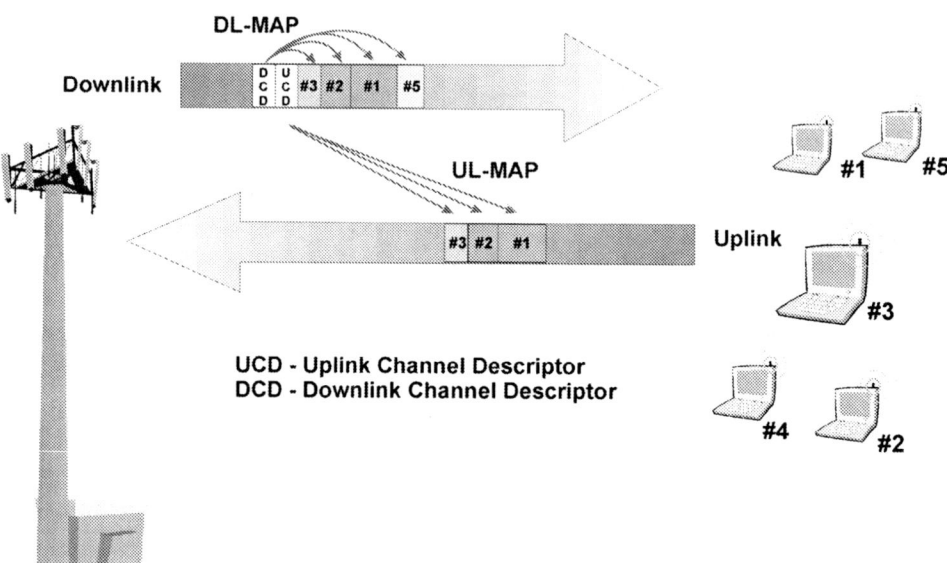

Figure 1.29, WiMax Channel Descriptors

Channel Coding

Channel coding is a process where one or more control and user data signals are combined with error protected or error correction information. The WiMAX system channel coding processes include error correction coding, interleaving and randomization.

Error Correction Coding

Error correcting codes are additional information elements that are sent along with a data signal that can be used to detect and possibly correct errors that occur during transmission and storage of the media. Error correction codes conform to specific rules or formulas to create the code from the data that is being sent. Error correction codes require an increase in the number of signal elements that are transmitted which increases the required data transmission rate. The WiMAX system can use a variety of error coding methods including Reed Soloman coding, convolutional coding (optional) and block turbo coding (optional).

Interleaving

Interleaving is the reordering of data that is to be transmitted so that consecutive bytes of data are distributed over a larger sequence of data to reduce the effect of burst errors. The use of interleaving greatly increases the ability of error protection codes to correct for burst errors. Many of the error protection coding processes can correct for small numbers of errors, but cannot correct for errors that occur in groups. The WiMAX system uses interleaving to map data onto non-adjacent sub-carriers to help overcome the effects of frequency selective (e.g. multi-path) distortion.

Randomization

Randomization is a process that rearranges data components in a serial bit sequence to statistically approximate a random sequence. For communication systems, randomization involves using a known randomization code or process in the transmitter and using the same code to decode the randomized sequence at the receiver.

The WiMAX system uses a pseudo-random binary sequence (PRBS) randomization process that ensures that there are no long sequences of bits that would cause the modulator to produce a high peak to average power ratio (PAPR) signal. PAPR is a comparison of the peak power detected over a period of sample time to the average power level that occurs over the same time period. A high PAPR would require the use of a more linear RF amplifier assembly increasing cost and decreasing power conversion efficiency (e.g. shorter battery life).

Duplex Transmission

Duplex transmission is the simultaneous transmission of two information signals that allows simultaneous (or near simultaneous) 2-way communication. The WiMAX system can use frequency division duplex (FDD), time division duplex (TDD) or half frequency division duplex (H-FDD).

FDD is the process of simultaneously allowing the transmission of information in both directions via separate frequency bands. When using FDD, each device transmits on one frequency while listening on a different one.

TDD refers to the process of allowing two way communications between two devices by time-sharing. When using TDD, device 1 transmits while device 2 listens for a short period of time. After the transmission is complete, the devices reverse their role so device 1 becomes a receiver and device 2 becomes a transmitter. The process continually repeats itself so data appears to flow in both directions simultaneously.

H-FDD is a process that allows for two-way communications between two devices through the combination of frequency division and time sharing. When using H-FDD, device 1 transmits on one frequency while the device 2 listens for a short period of time on that frequency. After the transmission is complete, the devices reverse their roles and device 2 transmits on a different frequency and the other device listens for a short period of time on that frequency. The process continually repeats itself so data appears to flow in both directions simultaneously. The use of H-FDD systems allows for the simplification of radio design as the transmitters and receivers in the same unit are separated in both frequency and time so a duplex filter is not required.

As shown in figure 1.30, the WiMAX system may use three types of duplexing: frequency division duplex (FDD), time division duplex (TDD) and half frequency division duplex (H-FDD). FDD allows the transmission of information in both directions at the same time by using separate frequency bands (frequency division). Half frequency division duplex uses different frequencies for transmission but does not allow transmission and reception at the same time.

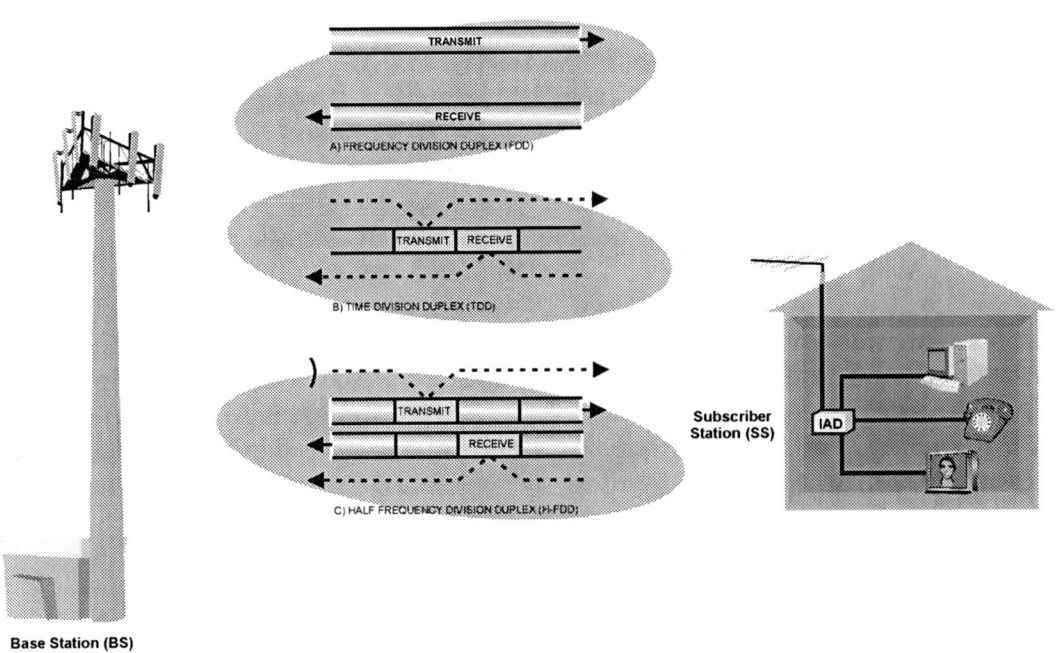

Figure 1.30, WiMax Duplex Transmission

When operating in time division duplex mode, WiMAX devices require reserved time periods to allow for transmission time (guard time) and to allow the device to transition between receive and transmit modes (transition gap).

Guard time is an amount of time that is allocated within a single time slot period in a communication system to help ensure variable amounts of transit times (e.g. from close and distant transmitters) do not cause overlap (collisions) between adjacent time slots. Transmission of information does not occur within the guard period.

Receive-transmit transition gap is the amount of time that is allocated between the reception of a packet and transmission of a packet in a time division duplex (TDD) system. Transmit-receive transition gap is the amount of time that is allocated between the transmission of a packet and the reception of a packet in a time division duplex (TDD) system.

WiMAX systems have the capability to dynamically change the amount of bandwidth that is transmitted in either direction through a process called adaptive time division duplex (ATDD). ATDD is a process of allowing two way communications between two devices by time sharing on the same communication channel (e.g. the same frequency) and the amount of transmission rate or time that is used by each device can dynamically change.

Figure 1.31 illustrates how the WiMAX system can use adaptive time division duplex transmission to vary the amount of bandwidth that is transferred in either direction. A base station initially sends data at a high rate. However, after the user has received the data, the base station begins to send a response at a high rate. The time periods allocated for transmission on the downlink and uplink continually vary to allow for variable data transmission rates.

Ranging (Dynamic Time Alignment)

Ranging is a dynamic time alignment process that allows a radio system base station to receive transmitted signals from mobile communication devices in an exact time slot, even though not all mobile communication devices are the same distance from the base station. Ranging keeps different mobile device

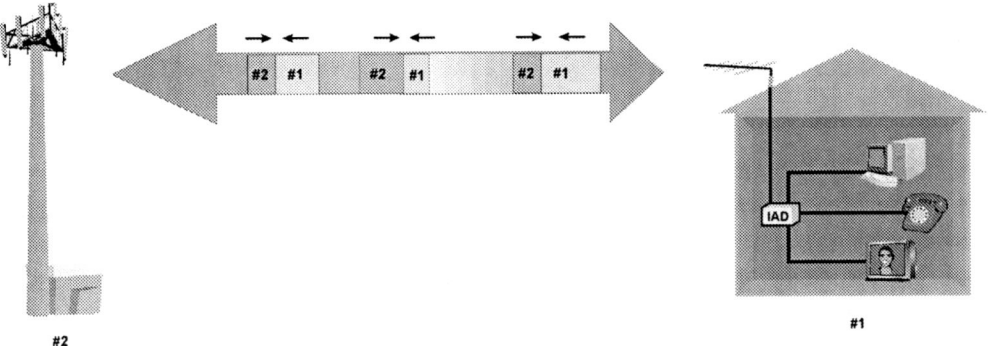

Figure 1.31, WiMax Adaptive Time Division Duplex (ATDD) Transmission

transmit bursts from colliding or overlapping. Ranging is necessary because subscriber stations may be moving or have been moved, and their radio waves' arrival time at the base station depends on their changing distance from the base station. The greater the distance, the more delay in the signal's arrival time. Transmission delay is approximately 3 microseconds per km (or 5 microseconds per mile). To perform time alignment, a subscriber station can advance or delay its transmission timing relative to the reference message that it receives on the downlink channel.

The WiMAX system uses two types of ranging: initial ranging and periodic ranging. During initial ranging, the WiMAX subscriber station transmits a brief ranging request message that allows the system to send back a ranging response message with the amount of timing offset that the subscriber station must use when it begins transmitting. After the subscriber station has attached to the system, the base station will continually send time alignment messages (periodic ranging) to the subscriber station to adjust (fine tune) its timing advance as it moves in the radio coverage area.

Initial ranging is the process of time aligning with a communication system before establishing a communication session. The initial ranging process involves synchronizing to an incoming transmission channel, sending a request at a particular time interval relative to the received channel and obtaining a response that allows for the time synchronization with the device or system.

Periodic ranging is the process of continuous time alignment with a communication system during a communication session. The periodic ranging process involves maintaining synchronization with an incoming transmission channel, periodically transmitting messages (such as sending user data) and receiving time alignment information from the other device or system.

If the subscriber station is continually communicating with the base station (receiving or transmitting data), the system can sense changes in the timing and send time alignment messages along with other packets of data to the subscriber station. If a significant amount of time has elapsed since a subscriber device has communicated with the system, the BS must initiate polling requests to reinitiate the ranging process.

Figure 1.32 shows how the relative transmitter timing in a subscriber station (relative to the received signal) is dynamically adjusted to account for the combined receive and transmit delays as the WiMAX radio is located at different distances from the base station antenna. In this example, the subscriber station uses a received burst to determine when its burst transmission should start. As the subscriber station moves away from the tower, the transmission time increases therefore causing the transmitted bursts to slip outside its time slot when it is received at the base station (possibly causing overlap to transmissions from other radios.) When the base station receiver detects the change in slot period reception, it sends commands to the subscriber station to advance its relative transmission time as it moves away from the base station and to be delayed as it moves closer.

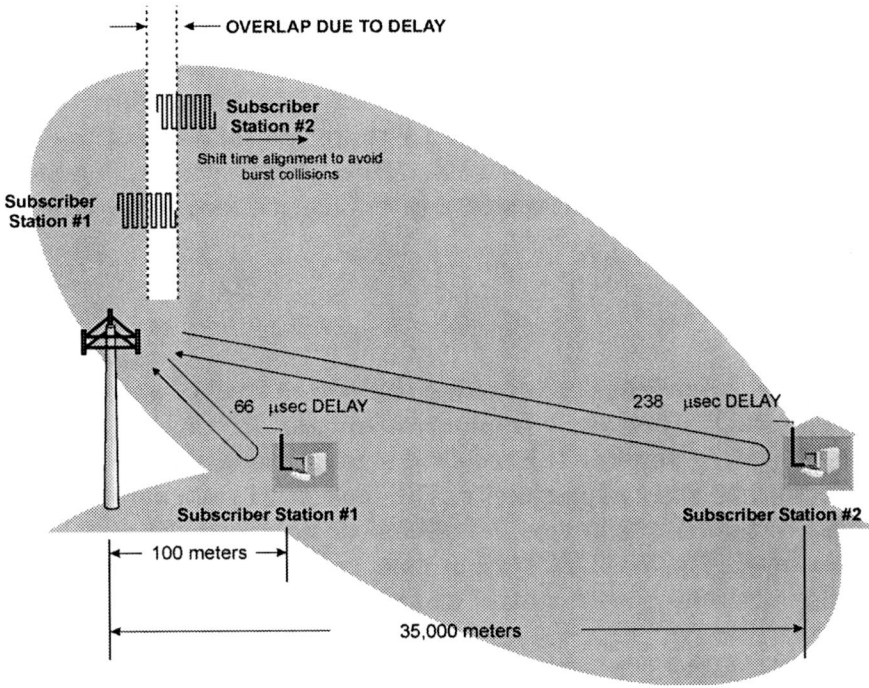

Figure 1.32, WiMax Ranging

Dynamic Frequency Selection (DFS)

Dynamic frequency selection is a process that allows devices or users to request, select or change an operating frequency at various times. Dynamic frequency selection involves sensing and assigning communication channels (such as radio frequencies) to transmitters (such as a radio base station) as they are required. The use of dynamic frequency selection in a WiMAX system allows for interference avoidance.

Interference avoidance is a process that adapts the access channel sharing method so that the transmission does not occur on specific frequency bandwidths. Using interference avoidance, devices that operate within the same frequency band and within the same physical area can detect the presence of each other and adjust their communication system to reduce the amount of interference caused by each other. This reduced level of interference increases the amount of successful transmissions therefore increases the efficiency and increases the overall data transmission rate. Dynamic frequency selection is used by WiMAX systems that operate in unlicensed frequency bands.

RF Power Control

RF power control is a process of adjusting the power level of a mobile radio as it moves closer and further away from a transmitter. RF power control is typically accomplished by sensing the received signal strength level to determine the necessary power level adjustments. The base stati then sends power control messages to the mobile device to increase or decrease the mobile device's output power level. The WiMAX system uses various forms of RF power control including open loop power control and closed loop power control.

Open loop power control is a process of adjusting the transmission power level for the subscriber station using the received power level. Open loop power control in the WiMAX system is used when the subscriber station initially attempts to accessing the system. When accessing the WiMAX system, the subscriber station calculates its' initial transmit power level using parameters that are broadcasted from the WiMAX system along with the signal strength that it receives. The smaller the signal strength it receives, the larger the initial transmit power level it uses when accessing the WiMAX system.

Closed loop power control is a process of adjusting the transmission power level for the mobile radio using the power level control commands from another transmitter that is receiving its signal (e.g. from a radio base station). The WiMAX system uses closed loop power control to continually adjust the transmitter level of the subscriber station as it moves, or as the signal level varies (such as when signal fading occurs or when obstructions move).

Figure 1.33 reveals how the radio signal power level output of a subscriber station is first determined by the received signal power level and is then adjusted by commands received from the base station to reduce the average transmitted power from the subscriber station. This lower power reduces interference to nearby cell sites and helps to ensure the signal level received by the base station from all the subscriber stations is approximately the same. As the subscriber station moves closer to the base station, less power is required from the subscriber station and it is commanded to reduce its transmitter output power level. The base station transmitter power level can also be reduced.

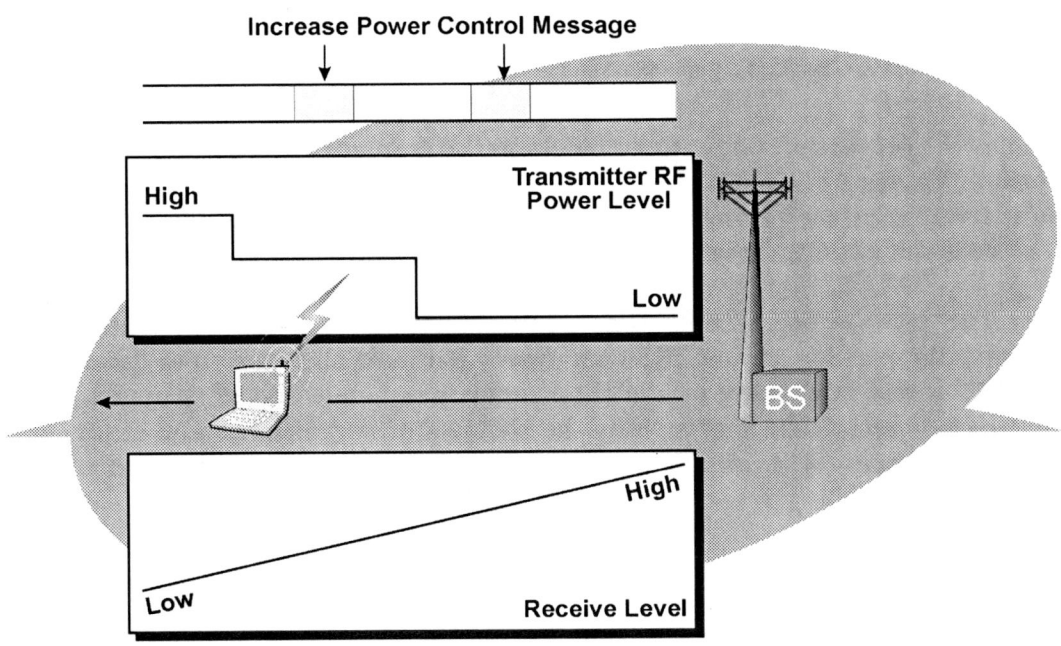

Figure 1.33, WiMax Power Control

Channel Measurement Reports

Channel measurement reports are groups of channel quality measurements that are sent from one radio device (such as a subscriber station) to another radio device (such as a system base station transceiver) that may be used to assist and adjust the radio transmission parameters (such as modulation type and RF power level) between the radio devices. Channel measurement reports are typically sent periodically to allow the base station or system to determine how the channel quality is changing to help it make decisions on channel modulation and coding types. The WiMAX system channel measurement reports include; radio signal strength indicators (RSSI), and channel to interference and noise ratio (CINR).

RSSI is the approximate level of a received signal captured by the radio device. Carrier to interference and noise ratio is a comparison of the information-carrying signal power to the interference and noise power in a system.

Payload Header Suppression (PHS)

PHS is the process of removing or blocking the transfer of packet header information. Payload header suppression is usually performed to remove redundant or unnecessary information such as a source and destination address that does not change for packets that are sent on a fixed communication link.

The PHS process begins with requesting a PHS session and negotiating the parameters or rules on how PHS will operate during the session to determine which bits and how many bits of the header may be altered or removed. PHS suppression operation may include the removal of address bits and other control information (such as IP address port number) that may be part of the header.

The header information is stored at the sending end and receiving ends to allow for the detection and removal of the header (the sending end) and the reinsertion of the header at the receiving end. The header is removed by using a payload header suppression mask (PHSM). The PHSM is a code or binary sequence that is used to allow, block, or modify specific bits in a header to create the information that will be transmitted. To allow the receiver to recreate information that is located within the header that sequentially changed, a payload header suppression index (PHSI) is used. The PHSI is an incremental value that is used to identify the sequence of payload header suppression messages.

When the compressed packet arrives at the receiver, the packet header information is recreated and inserted into the packet so the original data packet (header and data) is completely recreated.

Figure 1.34 shows how PHS can be used to increase the data transmission rate through a communication channel. An IP communication session that occurs over an unchanging (circuit switched) wireless data link can use PHS to increase the efficiency (higher data throughput) by removing redundant packet header data. When an IP session is setup over a circuit switched connection, the system first identifies that PHS will be used. The system then negotiates for which parts of the header will be changed or removed during transmission (establishes PHS rules). The negotiation associates (maps) the IP communication system to the data link connection and stores the unchanging information at each end of the communication link (the header mask). For each IP packet that is received, the IP address information and some control information is removed prior to transmitting the packet on the data link (packet compression). When the compressed packet arrives at the receiver,

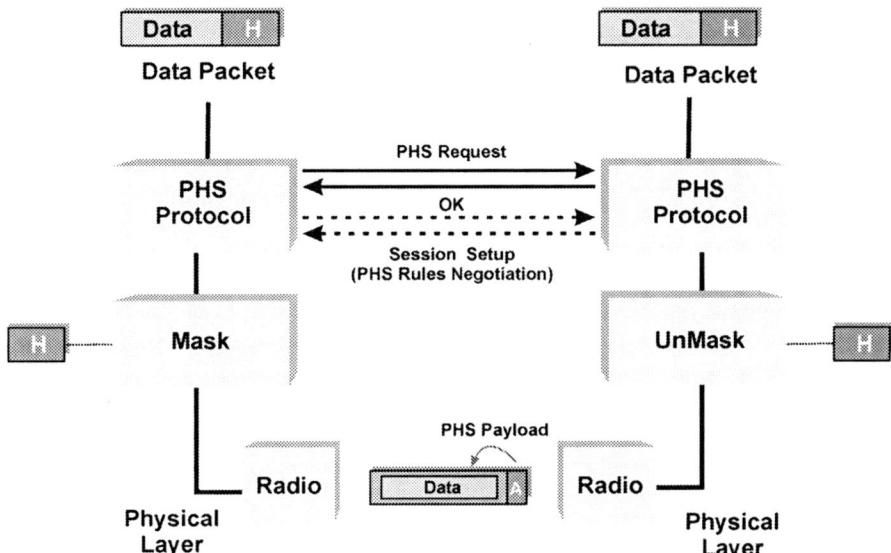

Figure 1.34, Payload Header Suppression

the IP address and packet header information is re-inserted on the packet so the IP data packet is completely recreated.

Convergence Sublayer (CS)

A convergence sublayer is a functional process within a communication device or system that adapts one or more transmission mediums (such as radio packet or circuit data transmission) to one or more alternative transmission formats (such as ATM or IP data transmission). The use of a convergence sublayer in the WiMAX system allows for the transparent flow of commands and media regardless of the media and type of transmission systems. The WiMAX system uses a convergence sublayer to convert IP, ATM and Ethernet protocols to the WiMAX system [5].

Sub Channelization (Sub-carriers)

Sub channelization is the dividing of communication channels into smaller sub-parts. The air interface portions of the WiMAX system divide a wide radio channel into several sub-carriers. A sub-carrier is a modulation signal that is imposed on another carrier that can be used to independently transfer information from other sub-carriers located on the radio channel. The WiMAX system sub-carrier signal types include a pilot sub-carrier (reference signal), guard sub-carrier (interference protection) and data sub-carrier (user information).

A pilot sub-carrier is a reference signal that serves as a control signal for use in the reception of other sub-carrier signals. A guard sub-carrier is one or more sub-carriers on a communication channel that is not used (null channel). The guard sub-carrier is dedicated to the protection of a communication channel from interference due to radio signal energy or time overlap of signals. Data sub-carriers are transmission channels that carry user information or data. Sub-carrier signals are referenced from the center of the radio channel. The sub-carrier that is located at the center of the radio channel is called the DC sub-carrier.

The number of sub-carriers in the WirelessMAN-OFDM system is 256. Of these, 55 are reserved as guard bands and 8 sub-carriers are used to transfer reference pilot signals. This allows up to 192 sub-carriers to be used for data transfer.

For a 20 MHz WiMAX OFDMA radio channel, there may be 2048 sub-carriers. Of these, approximately 70% can be used as data carriers, 25% are reserved to protect from interference (guard bands) and approximately 15% are used as reference signals (pilot channels) [6].

Figure 1.35 depicts how the WirelessMAN-OFDM system divides a wide radio channel into several independent orthogonal channels with smaller bandwidth. The sub-channel in the center of the RF channel is called the DC sub-

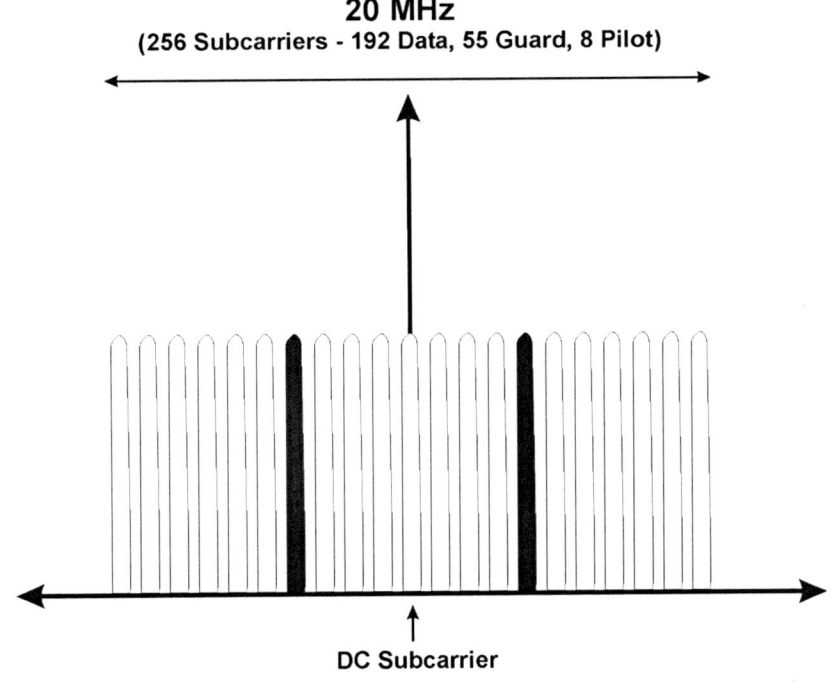

Figure 1.35, WiMax OFDMA Sub Channelization

carrier. Some of the sub-carriers are used as reference pilot channels and some are reserved for guard bands.

Retransmission Policy

Retransmission policy (automatic repeat request – ARQ) is the set of rules or processes used by networks to define if, when, and how retransmissions of data or information will occur. Some types of services (such as real time digital audio) do not use retransmissions because the delay for retransmission would take too long to offer any benefit.

Retransmission uses error detection, feedback, retransmission processes and the retransmission of blocks of data in packets. The WiMAX system uses two forms of ARQ; hybrid automatic repeat request (HARQ) and selective repeat (SR).

Selective Repeat (SR)

Selective repeat automatic repeat request is a data transmission control process that allows the receiver to request the retransmission of selective blocks of data.

Data to be transmitted is grouped into blocks and given a block sequence number (BSN). The maximum size of a data block in the WiMAX system is 2040 bytes. When blocks are transmitted using ARQ, each block is given a sub-packet identifier (SPID). An SPID is an index value that can be used to identify specific packets that are awaiting conformation in an automatic repeat request (ARQ) process.

As blocks are transferred between the sending device and the receiving device, acknowledgement messages are sent. WiMAX acknowledgement message types include selective, cumulative and cumulative with selective. ARQ messages can be sent as separate messages or the may be combined with other messages.

Hybrid Automatic Repeat Request (HARQ)

Hybrid automatic repeat request is a data transmission flow control process that uses a combination of the physical layer (PHY) and medium access control (MAC) layer to allow the receiver to stop and restart the retransmission of data over a transmission channel.

HARQ is a variation of a stop and wait ARQ. Stop and wait ARQ is a flow control process that allows the data flow to stop when packets are not received and wait until a successful retransmission is received before the data flow is restarted.

Physical RF Channels

WiMAX physical channels are the radio channels that are connected between transmitters and receivers. There are several types of physical RF channels that can be in the WiMAX system including single carrier, orthogonal carriers and multiple division access using orthogonal carriers.

Channel bandwidth is the difference between the upper frequency limit and lower frequency limit of allowable radio transmission energy for a channel. WiMAX systems can be setup to use different channel bandwidths dependent on the equipment and the objectives of the WiMAX system operator. Some of the key characteristics defined in system profiles include the channel bandwidth (e.g. 20, 25 or 28 MHz RF channel size).

Figure 1.36 shows that WiMAX radio channels can be single carrier or multiple carriers. The bandwidth of WiMAX radio channels can vary from 1.25 MHz to 28 MHz in steps of 1.75 MHz.

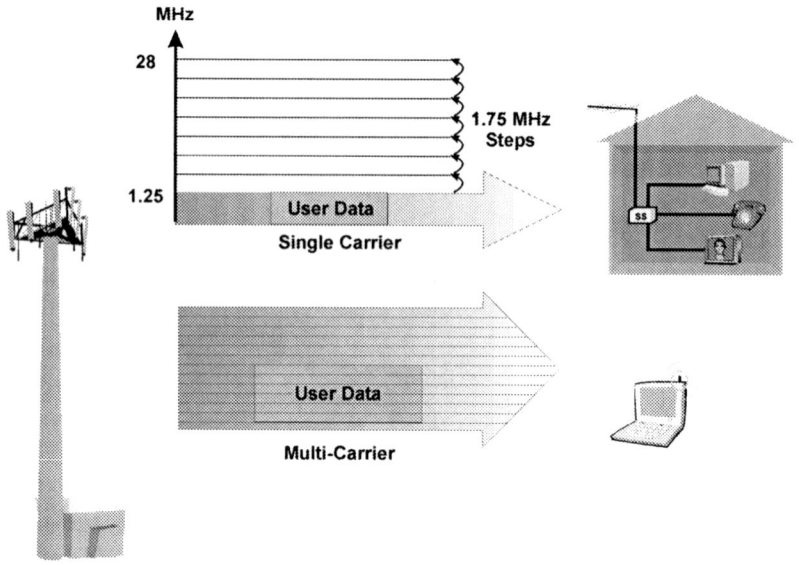

Figure 1.36, WiMax Radio Channels

Logical Channels

Logical channels are a portion of a physical communications channel that is used for a particular (logical) communications purpose. The WiMAX physical channel can have up to 65,535 logical channel connections and each connection can multiple service flows associated with it.

Connection ID (CID)

WiMAX logical channels are identified by a connection identifier (CID). A CID is a unique name or number that is used to identify a specific logical connection path in a communication system. Some connection channel IDs are reserved for control (management connection) and other connections are used for transporting user data.

Each type of connection has its own CID. A two-way connection requires two CIDs. For basic, primary and secondary connections, CID codes are assigned in pairs and they are the same for the downlink and uplink connections.

An initial ranging connection identifier is a code that is used during the initial connection to a wireless system to determine how much transmission timing adjustment is required. For WiMAX systems, the initial ranging CID is 0000 for standard transmission systems and FEFF for adaptive antenna systems.

A basic CID is a logical channel that is assigned during the initial ranging process. Basic CID connections are used for time sensitive MAC control messages such as RF power control and time alignment. The range of CIDs that can be assigned for basic CIDs is from 0001 to some number (m) selected by the operator.

A primary management CID is a logical channel that is used to transfer link control messages. The range of CIDs that are assigned as primary CIDs ranges from the address above the highest basic CID (m+1) to double the number of basic CIDs (2m).

A secondary management CID is a logical channel that is used for upper layer control messages such as DHCP and TFTP messages. The range of CIDs that is assigned as secondary management CIDs ranges from the address above the highest primary management CID (2m+1) up to connection ID FEFE.

A transport CID is a logical channel that is used to transfer user data. The range of CIDs that are assigned as transport CIDs ranges from the address above the highest primary management CID (2m+1) up to connection ID FEFE. Transport connections can use different CIDs in the uplink and downlink directions.

Multicast polling connection identifiers are used to prompt subscriber stations which are part of a multicast group that have data to transmit to attempt to transmit their data using a contention control process. The multicast polling CIDs range from FF00 to FFFC.

A broadcast connection identifier is used to transfer broadcast messages to all devices that are listening to the radio channel. The broadcast CID is FFFF.

Figure 1.37 outlines some of the CID codes that are used in the WiMAX system. The table shows that CID 0000 is reserved for initial ranging and the basic, primary, secondary and transport CIDs are dynamically assigned as needed. Other reserved CIDs include FEFF for adaptive antenna initial ranging, FF00 through FFFC for multicast polling, FFFD for fragmental broadcast messages, FFFE for padding messages and FFFF for broadcast messages.

Purpose	CID	Notes
Initial Ranging	0000	Initial Connection
Basic CID	0001 through m	Connection Setup
Primary Management	m+1 through 2m	Security and Control
Secondary Management or Transport Connections	2m+1 through FEFE	Upper layer control and user data transport
Adaptive Antenna Ranging	FEFF	Initial Connection
Multicast Polling	FF00 through FFFC	To control bandwidth requests
Fragmentable Broadcast	FFFD	
Padding	FFFE	
Broadcast	FFFF	Messages to all Subscriber Stations

Figure 1.37, WiMax CID Codes
Source: 802.16-2004, page 643

Service Flow ID (SFID)

A service flow identifier is a unique number that is assigned by a system that is used to identify the flow of a communication channel that is used for a specific service type. A WiMAX device may have multiple SFIDs per connection (per CID).

Figure 1.38 illustrates that the WiMAX system has logical connection and service flow channels. Each subscriber station has at least one connection channel and service flows may be assigned to the connections. In this diagram, a WiMAX base station has setup connections with 3 WiMAX subscriber sta-

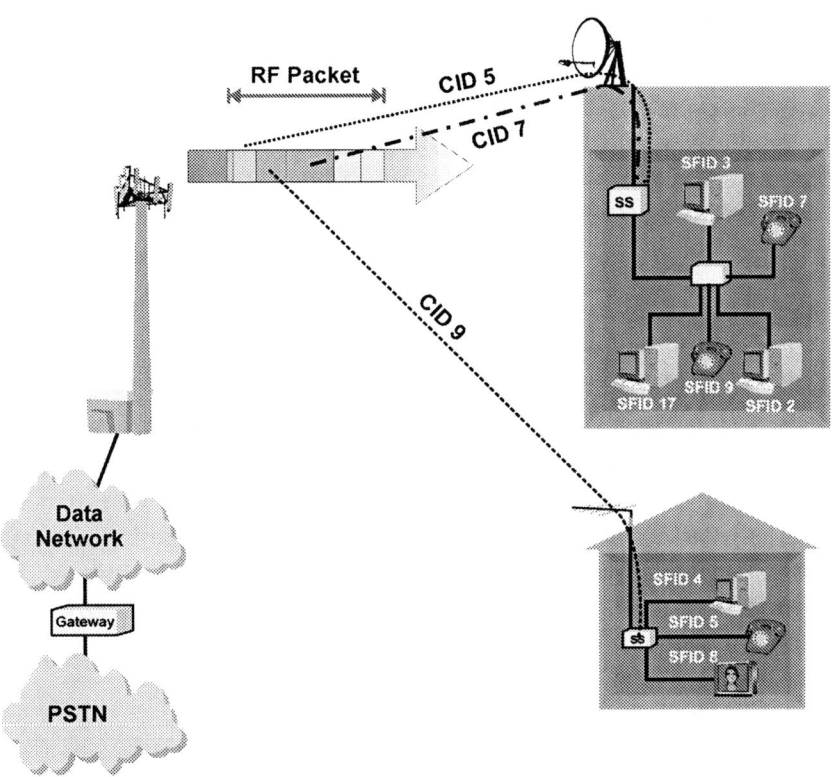

Figure 1.38, WiMax Logical Channels

tions. For home #1, the WiMAX transceiver connection is providing one type of service flow for an Internet web browser over a single connection. For home #2, the WiMAX base station has setup a single connection with two services flows; one for a web browsing computer and the other for an IP telephone. For the office user, the WiMAX base station has setup 3 connections on a single subscriber device (3 CIDs). Of these, 2 connections have 2 service flows providing web browsing and IP telephone service and the 3rd connection has a single service flow for web browsing service.

WiMAX Operation

The WiMAX system operates by coordinating the access to the radio channels and sending packets of data between base stations and subscriber stations. The basic operation of a WiMAX system involves channel acquisition, initial ranging, access control and radio link control.

Channel Acquisition

Channel acquisition is the process of finding and acquiring access to a communication channel. When WiMAX devices initialize (e.g. when they are turned on), they begin a channel scanning process. Channel scanning is the process of searching through multiple radio channels to find signals that indicate a channel is available on which to communicate. The WiMAX device will typically have a stored list of frequency channels for it to scan in order to reduce the amount of scanning time. These frequency channels may be preprogrammed by or for a WiMAX system operator so the WiMAX device will initially try to connect to a specific WiMAX system.

When the WiMAX device has found one or more WiMAX radio channels, the device will decode the channel and look for packets of data that have a frame control header that contains a downlink channel description (DCD) message and an uplink channel description (UCD) message. The DCD message contains parameters that are necessary or that will assist it to access the device in receiving information from the downlink channel on the communication system. The UCD message provides the device with the parameters that are necessary to access the communication system.

Initial Ranging

Initial ranging is the process of estimating the distance or propagation time between a transmitter and receiver. Ranging information may be used to assist in the establishment of operating parameters for the transmitter and receiver. The transmitter power level and packet transmission delay time ensure packets do not overlap with transmission from other devices.

During the initial ranging process, the base station is assigned the basic CID that will be used to control the radio operations of the subscriber device. After the basic CID is assigned, a primary management CID may be assigned to allow for authentication and the establishment of other CID channels. A secondary CID may be assigned to allow the downloading of configuration files and the assignment of an IP address using dynamic host configuration protocol (DHCP).

Figure 1.39 depicts the basic channel acquisition processes that may be used in the WiMAX system. The subscriber station begins by scanning a set of potential WiMAX frequencies. If it finds a WiMAX radio channel, it synchro-

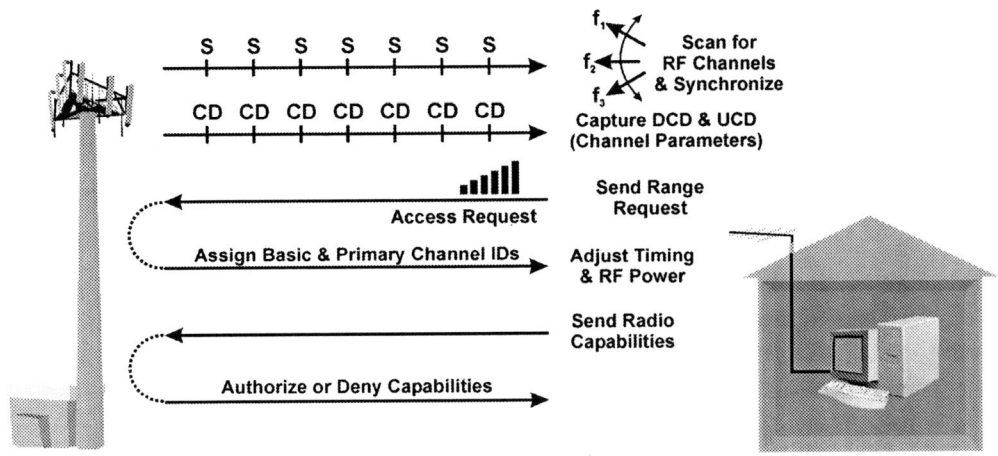

Figure 1.39, 802.16 Channel Acquisition and Initial Ranging

nizes with the RF channel and acquires the downlink channel descriptor (DCD) and uplink channel descriptor (UCD) messages to determine how to access the system. The subscriber station then sends initial ranging request messages to get the attention of the system and to receive timing adjustment information. This process starts by transmitting at a lower RF power level and gradually increasing until the system responds with an assignment of basic and primary control identifiers (CID). The subscriber station then sends its transmission capabilities to the base station and the WiMAX system responds with an authorization or denial of service for these transmission capabilities.

Medium Access Control

Medium access control is the process used by communication devices to gain access to a shared communications medium or channel. The methods for controlling access to WiMAX systems may be assigned ("non-contention based") or random ("contention based").

When the WiMAX system uses contention free access control the subscriber station must wait for polling messages before responding. If contention based access control is used (e.g. best effort service), the subscriber device must compete for access to send its packets. The WiMAX system can mix contention free and contention based access on the same radio channel.

Contention free access is provided by defining time periods that specific devices will use when communicating with the system. Because all the devices listening to the WiMAX radio channel can hear these messages, devices will not transmit during the assigned time periods.

Contention based access is provided through the use of contention slots and the collision sense multiple access (CSMA) process. The WiMAX channel descriptors define specific time periods ("contention slots") that contention based WiMAX devices must use when accessing the WiMAX system. Contention slots are dedicated time intervals (time slots) on a communication channel that can be used to allow devices to randomly request service from a system.

When contention based WiMAX subscriber stations access the system, they first obtain the contention time slot interval and the system access parameters (e.g. initial access transmit power level). After the contention slot time period has started, the subscriber station begins to transmit an access message at a low RF power level. If the subscriber station hears a positive response to its access request message, it can transmit its package. If the subscriber device does not hear a response (e.g. another device has transmitted at the same time), it must stop transmitting and wait a random amount of time before attempting to access the system again. Each time the device attempts to access the system and fails, it must wait a longer amount of time before attempting to access the system again. This prevents the possibility of many collisions between devices that are attempting to access the system at approximately the same time.

Figure 1.40 illustrates how the WiMAX system can mix contention free and contention based access control on a WiMAX radio channel. This diagram shows that the downlink channel contains downlink and uplink descriptor messages that define when subscriber stations are allowed to transmit. For

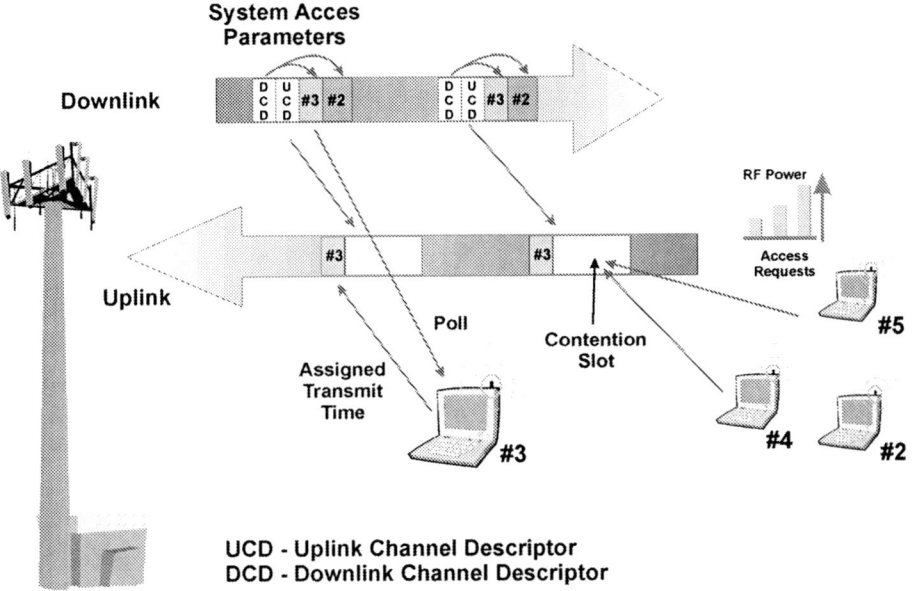

Figure 1.40, WiMax Access Control

unicast polled devices (contention free), they are assigned specific time periods to transmit from a polling message. For multicast polled, broadcast polled or best effort devices (contention based), they compete during the contention time slot periods.

Radio Link Control (RLC)

Radio link control protocol is a layer 2 (link layer) that is used to coordinate the overall flow of data packets across the radio link. RLC uses error detection and data retransmission to increase the reliability of the radio link while reducing the error rate. WiMAX radio link control functions include power level control, periodic ranging, burst profile changes and bandwidth requests.

Power control is the process of adjusting the power level in a wireless system where the base station receiver monitors the received signal strength of mobile radios. Control messages are transmitted from the base station to the mobile telephone commanding it to raise and lower its transmitter power level as necessary to maintain a good radio communications link.

Ranging may need to be performed after the subscriber station has been inactive for a while. A timer (the T4 timer) that is continuously reset as the subscriber station communicates with the system helps determine this. If the subscriber station (SS) has not communicated with the system in awhile, the timer will not be reset and it will expire. If the timer expires, the SS must again perform ranging with the system.

The base station is responsible for assigning burst profiles. However, the subscriber station may request changes to the burst profile. This may occur as a result of an increase in the bit error rate of the received signal due to fading or interference. The subscriber station may request a change in burst profile that is more robust or offers a higher data transmission rate. The base station may grant the request, negotiate parameters or reject the request.

During a WiMAX communication session, changes in bandwidth may be requested. The subscriber station may send bandwidth request messages to the base station to increase or decrease its allocated bandwidth. Bandwidth request messages may be sent as independent messages or they may be piggybacked with other messages.

Types of Connections

Connection types are the purposes (such as for control or for transfer of user data) and characteristics (such as packet or dedicated circuits) of physical and/or logical communication paths. The types of connections in the WiMAX system include a basic connection, primary management connection and a transport connection.

Basic Connection

A basic connection is a WiMAX control channel that performs basic link management functions such as setting up, changing physical characteristics (e.g. power level and timing) and terminating radio connections. All WiMAX connections must have a basic connection.

Primary Management Connection

A primary management connection is used to transport messages that are less time sensitive. They are typically used for higher level functions such as authentication and the setup of logical channels.

Secondary Management Connection

A secondary management connection is used to transport messages that perform higher level functions such as IP address assignment (DHCP), equipment configuration control (SNMP) and file transfer (TFTP). Secondary management connections are optional.

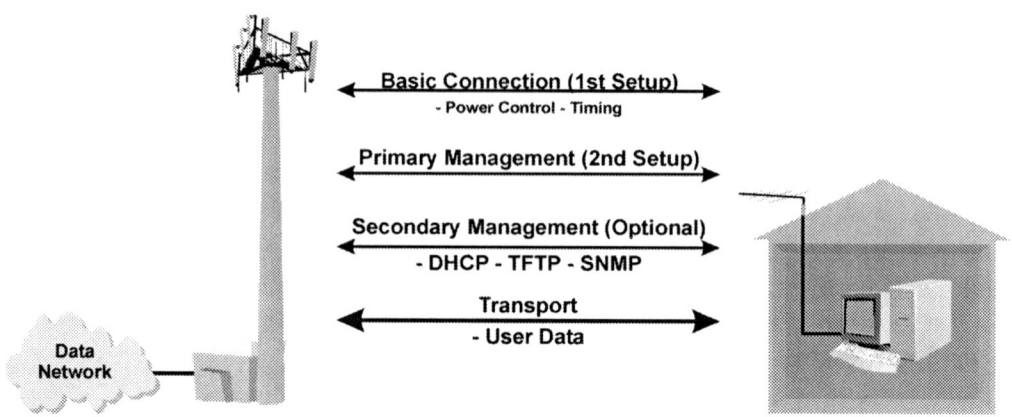

Figure 1.41, WiMax Connection Types

Transport Connection

A transport connection is a unidirectional connection that is used to transport user data. Each transport connection has unique QoS parameters associated with it. Transport connections are typically assigned in pairs (uplink and downlink). Figure 1.41 shows the different types of connections used in a WiMAX system.

Quality of Service (QoS)

QoS is one or more measurements of desired performance and priorities of a communications system. The WiMAX system is designed with the ability to apply different QoS levels to downlink and uplink connections as well as provide multiple service types on a single connection to each user. QoS measures for WiMAX systems may include service availability, data throughput, delay, jitter, and error rate.

Service Availability

Service availability is the ratio of the amount of time an authorized user is able to access the services compared to the total time service is supposed to be available. Service availability can be affected by a variety of factors including admission control and oversubscription.

Admission control is the process of reviewing the service authorization level associated with users and determining the extent to which network resources will be allocated if they are available. Admission control is used to adjust, limit or assign the use of limited network resources to specific types or individual users. Admission control may allow for the assignment of higher access level priority for specific types of users such as public safety users.

Oversubscription is a situation that occurs when a service provider sells more capacity to end customers than a communications network can provide at a specific time period. This provides a benefit of reduced network equipment and operational cost.

Oversubscription is a common practice in communications networks as customers do not continuously use the maximum capacity assigned to them and they access the networks at different time periods. Unfortunately, over-subscription in telecommunications can cause problems when customers do attempt to access the network at the same time. For example, when customers open their presents at a holiday event (e.g. Christmas) and attempt to access the Internet at the same time.

Data Throughput

Data throughput is the amount of data information that can be transferred through a communication channel or transfer through a point on a communication system. Different types of applications and services require different data throughput rates. WiMAX systems can be setup to provide a variety of data services that can have controlled data throughput rates. This includes the ability to provide constant data throughput rates for committed bit rate (CBR) services (e.g. T1 or E1 service) and to best effort services (e.g. for residential broadband access service).

Delay

Delay is the amount of time it takes for a signal to transfer or for the time that is required to establish a communication path or circuit. WiMAX systems can use scheduling servers to prioritize packets and control the maximum amount of delay for specific users or groups of users.

Jitter

Jitter is a small, rapid variation in arrival time of a substantially periodic pulse waveform resulting typically from fluctuations in the transmission system. While jitter may not be able to be directly controlled, it can be minimized by using channels that have larger bandwidths (higher data throughputs). This will minimize both delay and the amount of delay variation (jitter).

Error Rate

Error rate is a ratio between an amount of information that is received in error as compared to the total amount of information that is received over a period of time. Error rate may be expressed in the number of bits that are received in error on the number of blocks of data (packets) that are lost over a period of time. WiMAX error rates can be affected by a number of factors including signal quality and system configuration. Some of the common error rate measures for WiMAX include bit error rate (BER) and packet loss rate (PLR).

Bit Error Rate (BER)

BER is calculated by dividing the number of bits received in error by the total number of bits transmitted. It is generally used to denote the quality of a digital transmission channel. Bit errors can occur randomly over time (random errors) or in group (burst errors).

Random errors are bits in a received digital signal that are received in error that occur in such a way that each error can be considered statistically independent from any other error. Burst errors are the distortion or failure of a digital receiver to correctly decode groups of digital bits. Burst errors typically have a high bit error ratio (BER) compared to the overall BER of a communication link or channel.

Packet Loss Rate (PLR)

Packet loss rate is a ratio of the number of data packets that have been lost in transmission compared to the total number of packets that have been transmitted. Some applications (such as digital television) are more sensitive to packet loss rate than bit error rates.

Scheduling Services

Scheduling services are the medium access control functions (data flow control) that define how and when devices will receive and transmit on a communication system. The types of services that WiMAX can provide range from guaranteed bandwidth with low delay unsolicited grant service (UGS) to random access best effort (BE) service. WiMAX systems use a grant management system to coordinate the request for new services and changes to existing services (such as requesting more bandwidth). The WiMAX system uses a combination of time division multiple access, polling and contention based flow control to provide specific types of services to users.

Time division multiple access (TDMA) is a process of sharing a single radio channel by dividing the channel into time slots that are shared between simultaneous users of the radio channel. When a subscriber communicates on a WiMAX system using TDMA, he/she is assigned a specific time position on the radio channel. By allowing several users to use different time positions (time slots) on a single radio channel, TDMA systems can guarantee a constant data rate with a minimal amount of flow control overhead.

Polling is the process of sending a request message (usually periodically) for the purpose of collecting events or information from a network device. The receipt of a polling message by a device starts an information transfer operation for a specific time period. Polling may be performed with specific units (unicast), to groups of units (multicast) or to all units (broadcast).

Unicast polls are requests for data transmission or responses to commands that are only sent between a sender (polling device) and receiver (polled device). When a subscriber station is responding to a unicast polling message, no other devices are allowed to transmit.

Multicast polls are requests for data transmission or responses to commands that are sent from a polling device to several receiving devices which are part of a multicast group. When a device receives a multicast polling message for its group, it will respond if it has data to send. When a subscriber station is responding to a multicast polling message, others may also have information to transmit. For multicast poll messages, subscriber stations must use contention based access (on the contention slot) to send their data.

Broadcast polls are requests for data transmission or responses to commands that are sent from a polling device to all devices that are able to receive its broadcasted polling message. When a device receives a broadcast polling message, it will respond if it has data to send. For broadcast poll messages, subscriber stations must use contention based access (on the contention slot) to send their data.

The amount of time between polling messages is called the polling cycle. The time between polling cycles is a balance between delay (more polling messages is less delay) and overhead (more polling messages increases the percentage of data that is used for control messages).

Figure 1.42 illustrates the different types of polling that are used in the WiMAX system. A device that is part of a multicast group, has received a multicast polling message, must compete for access to send its data. Finally, for a broadcast polling message, any device that has data will compete for access to send its data.

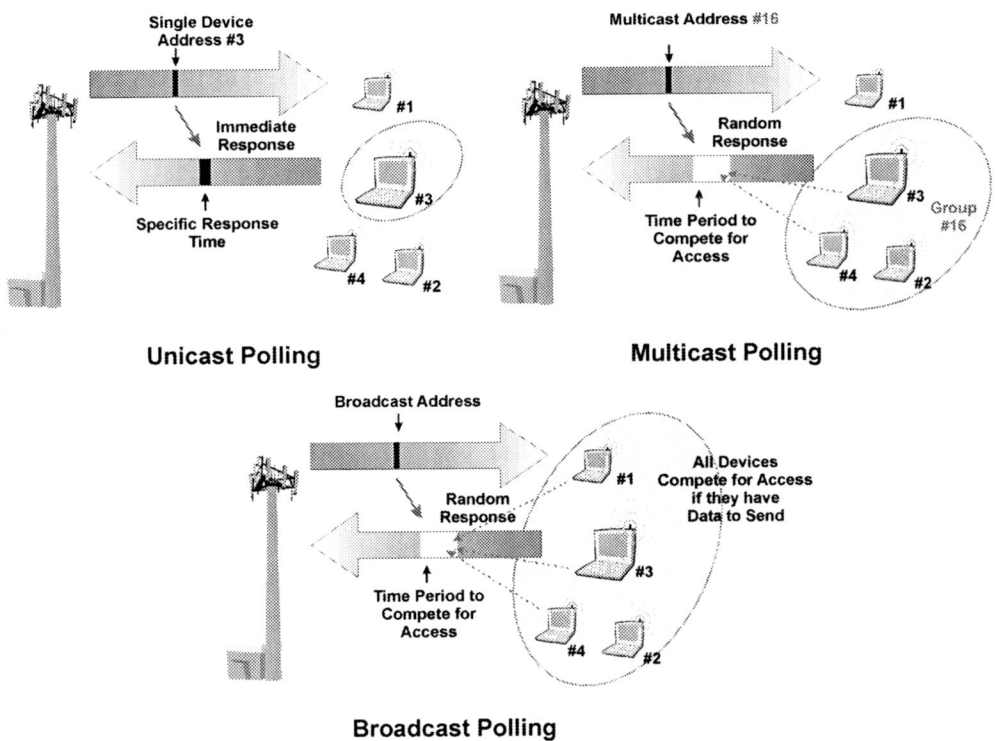

Figure 1.42, WiMax Polling Types

Contention based access control is the independent operation (distributed access control) of communication devices (stations). In a contention-based system, communication devices randomly request service from channels within a communication system. Because communication requests occur randomly, two or more communication devices may request service simultaneously. The access control portion of a contention based session usually involves requiring the communication device to sense for activity before transmitting and listening for message collisions after sending its service request. If the requesting device does not hear a response to its request, it will wait a random amount of time before repeating the access attempt. The amount of time waited between retransmission requests increases each time a collision occurs.

The WiMAX system defines time periods that subscriber stations can use for contention based access. When subscriber units desire to initiate requests to the system that are not scheduled from a polling message, they must access the process during the contention time slots period. The contention time slot period is periodically broadcast on the downlink channel along with other channel access control information.

Figure 1.43 shows how contention based access control can be performed on a WiMAX system. Channel descriptors are periodically broadcasted on the

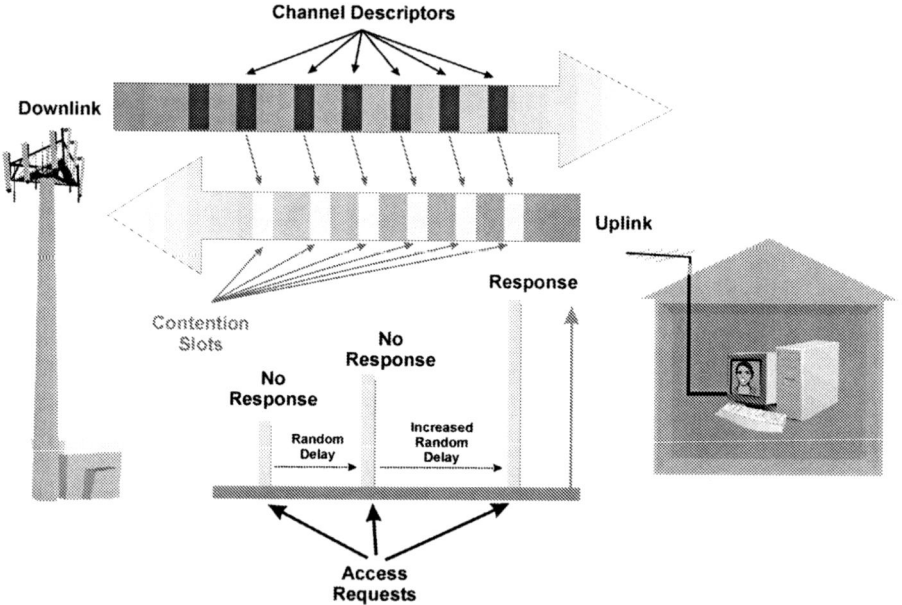

Figure 1.43, WiMax Contention Based Access Control

downlink radio channel that provides the time intervals for the contention slots. Subscriber devices that use contention based access must compete during these time periods. The WiMAX subscriber station will initially attempt to access the system at a relatively low power level. If the subscriber station does not hear a response to its request, it will wait a random amount of time, increase its transmitted power level and attempt access again. The subscriber station will continue to wait increasing amounts of time each time and increases its transmitted power level each time an access attempt fails until it receives a response from the system.

The WiMAX system uses a grant management process for the requesting and allocation (granting) of resources (such as transmission time or bandwidth). Subscriber stations can request changes to the type of services they require (e.g. increases or decreases in bandwidth) by transmitting a bandwidth request header and the system can decide to grant, adjust or not authorize the grant request.

The WiMAX system can grant resources based on a connection or based on a specific subscriber station. A grant per subscriber station is the allocation of transmission bandwidth that affects the transmission for all the connections associated with a subscriber station. A grant per connection is the assignment of bandwidth which only affects the transmission for a specific connection on a subscriber device.

Bandwidth requests can be in aggregate or incremental form. An aggregate request is a message that defines the amount of a resource (such as transmission bandwidth) that is requested to provide for a combined group of applications or services. An incremental request is a message that defines the additional amount of a resource (such as transmission bandwidth) that is requested to provide for an application or service. Bandwidth request messages may be sent as stand alone messages or they may be piggybacked in the payload of another packet of data.

Unsolicited Grant Service (UGS)

Unsolicited grant service is a service flow in which the transmission system automatically and periodically provides a defined number of timeslots and fixed packet size that is used by a particular receiver. UGS is commonly used to provide services that require a constant bit rate (CBR) such as audio streaming or leased line (e.g. T1 or E1) circuit emulation.

UGS provides a constant bit rate for a single connection. A subscriber device may need additional bandwidth for an additional service that is added to a connection or to temporarily provide more bandwidth on the UGS connection. To request more bandwidth on a UGS connection, a poll me bit or slip indicator bit may be used.

A poll me bit is a signaling message in a data field within the header of a data packet that indicates that the device would like to be polled. The poll me bit indicates to the base station that the subscriber device needs to be polled for a service other than for the current UGS service.

For transmission to synchronous connections, timing inaccuracies may result in the need to transfer additional bits if the clock of one connection is slightly faster than the other connection. When the buffer of the faster connection indicates the number of bits to be transmitted may soon run out, a slip indicator bit may be used. The slip indicator is a signaling message within the header of a data packet that indicates that the data transmission queue of that device is changing (slipping) and that the device needs more bandwidth to keep up with the transmission queue. This allows the base station to temporarily assign additional bandwidth until the transmission buffer has caught up.

Figure 1.44 shows how WiMAX unsolicited grant service (UGS) operates. Subscriber stations are assigned to receive and transmit during assigned time intervals. The subscriber station may use the poll me bit in the header to indicate it wants to be polled so it can send data for another service. When the base station receives the poll me bit, it sends a polling message which allows the subscriber station to send a packet of data that is independent of the UGS packets.

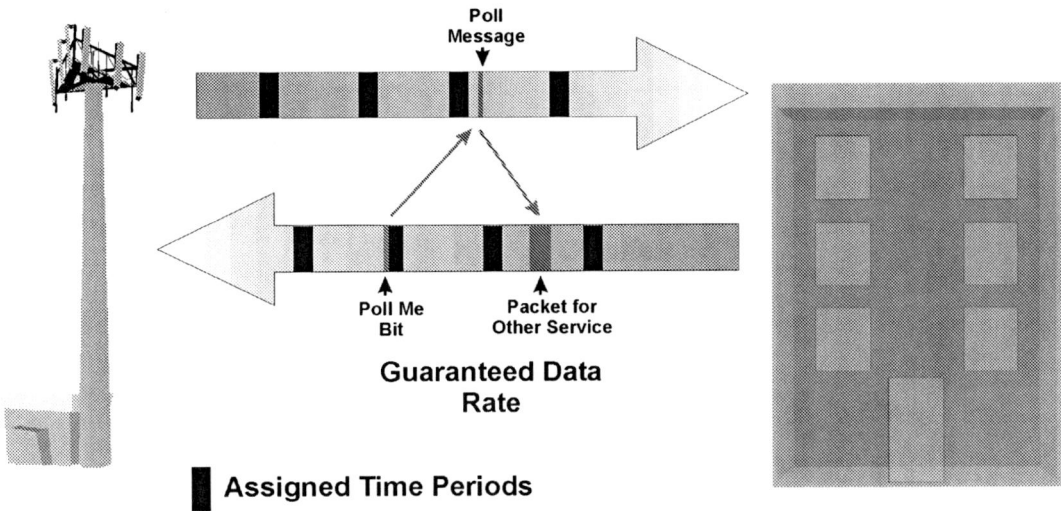

Figure 1.44, Wireless Unsolicited Grant Service (UGS)

Real Time Polling Service (RTPS)

RTPS is the periodic sending of polling messages that allow the subscriber station to regularly request additional allocations of bandwidth and coding types. For RTPS, the subscriber station can receive and send variable length packets. RTPS could be used for applications such as digital video which benefit from variable transmission rates.

Figure 1.45 shows WiMAX real time polling service. For RTPS, subscriber stations regularly receive polling messages that allow them to send data. When the subscriber station receives unicast (single user) polling message, it can immediately send packets back to the system.

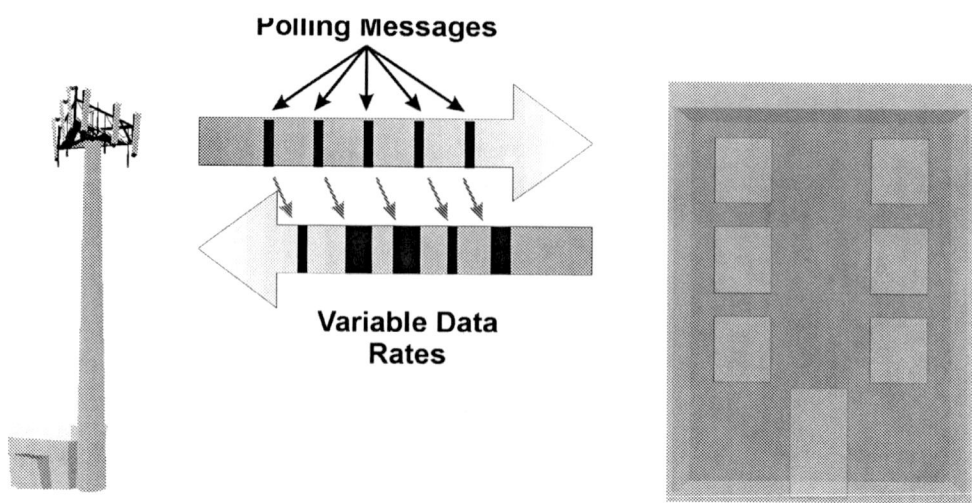

Figure 1.45, Wireless Real Time Polling Service (RTPS)

Non-Real Time Polling Service (nRTPS)

nRTPS is the random, unscheduled sending and receiving of packets through the use of polling messages and possibly contention based access control. While the system can send polling messages, the subscriber station may be also allowed to randomly compete for access when it wants to obtain bandwidth. For nRTPS, a base station typically polls subscriber approximately every second. nRTPS is used for delay tolerant applications that can use variable data transfer rates.

Figure 1.46 shows WiMAX non-real time polling service. This diagram shows that for nRTPS, subscriber stations periodically receive polling messages that allow them to send data. However, subscriber stations may also randomly compete for access when they desire to transmit information.

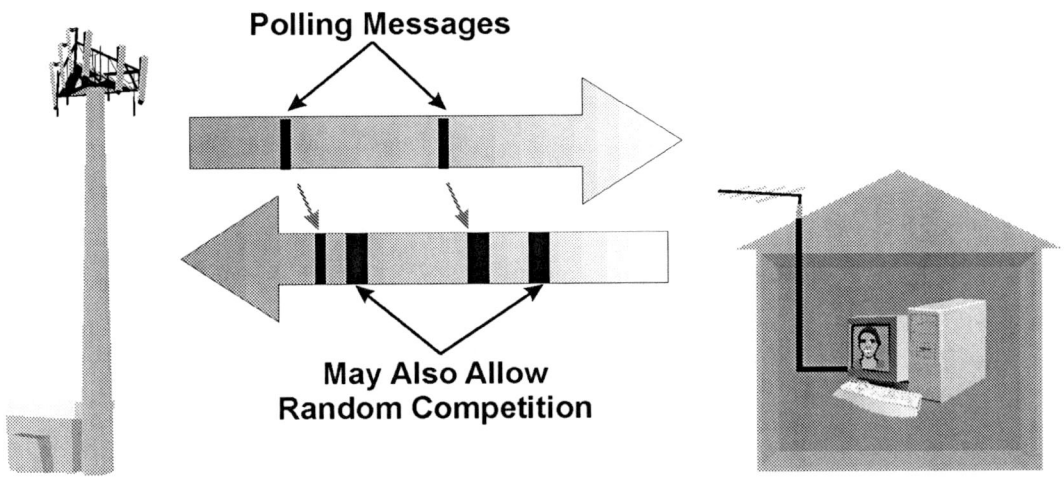

Figure 1.46, Wireless Non-Real Time Polling Service (nRTPS)

Best Effort Service (BE)

Best effort is a level of service in a communications system that doesn't have a guaranteed quality of service (QoS). For BE service, the subscriber station communicates with the WiMAX system when it desires to transmit or to obtain more bandwidth. Best effort service could be used for residential broadband (Internet browsing) applications.

Figure 1.47 shows WiMAX best effort service. For BE service, subscriber stations randomly compete for access when they desire to transmit information.

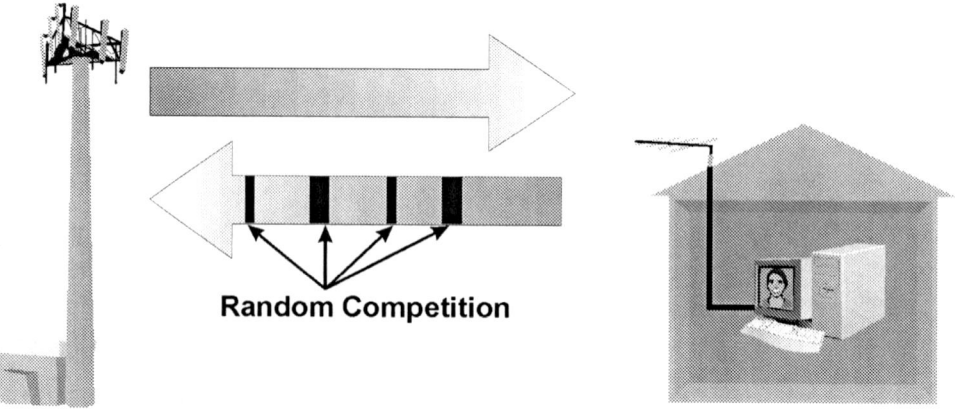

Random Competition

Figure 1.47, Wireless Best Effort Service

Service Flows and Classes

WiMAX services are the providing of information transfer to or between users. Information transfer services can have a variety of characteristics that can be selected and varied by the operator. The WiMAX system uses services flows to identify specific transmission characteristics for specific user services and a single user may have multiple service flows. Common sets of service characteristics may be defined by service classes.

Service Flows

Service flows are communication channels (e.g. a stream of packets) that have particular service characteristics associated with the transfer of data. For example, a communication link might have several service flows associated with it; a real time service flow for voice communication, a high-integrity service flow (low error rate) for data file transfer and a best effort service flow for Internet web browsing.

Each service provided on a WiMAX system is associated with a service flow. Service flows are requested, established and ended. When subscriber stations request services, the system can evaluate and negotiate the requested characteristics at any time.

A single WiMAX subscriber station may have multiple service flows for each connection (service flows can be different in different directions). Service flows can be dynamically added, changed and ended.

Service flows are uniquely identified by a service flow identifier (SFID). A SFID is associated with a specific connection identifier to determine the service characteristics a specific user will receive on that particular device.

Service Class

A service class is a set of communication parameters that are used or assigned to provide transmission flows that can provide services that meet specific quality of service (QoS) requirements. The WiMAX system can use service names (a label or a tag) to identify a particular type of service flow to provision for a particular user or device.

Services may be authorized, admitted or active. The creation of services begins with the determination of services that the user or device has been authorized to use. This is followed by the setup of network equipment to enable the transfer or processing of information. When the user begins to transfer and/or process information, the service becomes active.

WiMAX Certification

WiMAX certification is the process of testing products to ensure they conform to the industry standards and are interoperable with other WiMAX products. The certification of products is important because it provides consumers and businesses with confidence that the wireless WiMAX products will perform correctly and will interoperate with other WiMAX devices.

WiMAX Forum

TheWiMAX forum is a non-profit industry group that was created to assist in the development, certification and promotion of IEEE 802.16 and ETSI HiperMAN standards.

Certification Process

The WiMAX certification process is the steps, processes and tests which are performed to ensure that products will perform as desired and will reliably interoperate with other devices that are certified.

Testing for WiMAX products can be performed using testing conformance standards. The WiMAX test standards include a protocol implementation conformance statement (PICS), a test suite structure (TSS) and a radio conformance testing (RCT).

A protocol implementation conformance statement proforma is a document that is provided by a company or testing facility that states that the product or system provides and supports a specific set of commands and protocols. The 802.16 system has many capabilities and WiMAX devices typically are designed to support only a limited set of the protocols and capabilities.

A test suite structure (TSS) is a set of testing equipment configurations and procedures that are used to evaluate the operation and performance of products or systems. For the WiMAX system, the TSS performs operational testing of key functions of WiMAX devices including connecting radio links, authenticating, setting up, and changing services.

Radio conformance tests are a set of procedures that are used to evaluate the operation and performance of the radio part of wireless products or systems.

References:

1. 802.16 WiMAX industry standard, 2004, www.IEEE.org, page 2.

2. i-Max "Technical Information," www.WiMAXforum.org/tech. 2 April, 2004.

3. WiMAX Forum, "WiMAX's technology for LOS and NLOS environments," www.WiMAXForum.org.

4. David Johnston and Hassan Yaghoobi, "Peering Into the WiMAX Spec," CommsDesign, www.commsdesign.com, 21 Jan, 2004.

5. 802.16 WiMAX industry standard page 17.

6. "Scalable OFDMA Physical Layer in IEEE 802.16 WirelessMAN," www.Intel.com, 20 August, 2004.

Appendix 1

Acronyms

3DES-Triple Data Encryption Standard
AAS-Adaptive Antenna System
ABR-Available Bit Rate
ACK-Acknowledgment
AISN-ARQ Identifier sequence number
AK-Authorization Key
AMC-Adaptive Modulation and Coding
AOA-Angle of Arrival
AOD-Angle of Departure
AP-Access Point
ARQ-Automatic Repeat Request
ASIC-Application Specific Integrated Circuit
ATDD-Adaptive Time Division Duplex
ATM-Asynchronous Transfer Mode
ATS-Abstract Test Suite
BE-Best Effort Service
BER-Bit Error Rate
BPSK-Binary Phase-Shift Keying
BRAN-Broadband Radio Access Networks
BRS-Broadband Radio Service
BS-Base Station
BSN-Block Sequence Number
BTC-Block Turbo Code
BWA-Broadband Wireless Access
BW-Bandwidth
C/I-Carrier To Interference Signal Ratio
CAC-Call Admission Control
CBC-Cipher Block Chaining
CBLER-Coded Block Error Rate
CBR-Constant Bit Rate
CC-Convolutional Coding
CDMA-Code Division Multiple Access
CID-Connection Identifier

CINR-Carrier to interference-plus-noise ratio
CLR-Cell Loss Rate
COS-Class Of Service
CPE-Customer Premises Equipment
CRC-Cyclic Redundancy Check
CSCH-Centralized Scheduling
CS-Convergence Sublayer
CSMA/CA-Carrier Sense Multiple Access/Collision Avoidance
CTS-Clear To Send
DAMA-Demand Assigned Multiple Access
DBPC-Downlink Burst Profile Change
DCD-Downlink Channel Descriptor
DCF-Distributed Coordination Function
DES-Data Encryption Standard
DFS-Dynamic Frequency Selection
DHCP-Dynamic Host Configuration Protocol
DIUC-Downlink Interval Usage Code
DL-Downlink
DLFP-Downlink Frame Prefix
DOCSIS+-Data Over Cable Service Interface Specification +
DSA-Dynamic Service Addition
DSC-Dynamic Service Change
DSCH-Distributed Scheduling
DSCP-Differentiated Service Code Point
DSD-Dynamic Service Delete
DSL-Digital Subscriber Line
DVB-Digital Video Broadcast
EAP-Extensible Authentication Protocol
EGC-Equal Gain Combiner
EIRP-Effective Isotropic Radiated Power

EKS-Encryption Key Sequency
FC-Fragment Control Field
FDD-Frequency Division Duplex
FDMA-Frequency Division Multiple Access
FDM-Frequency Division Multiplexing
FEC-Forward Error Correction
FFT-Fast Fourier Transform
FHSS-Frequency Hopping Spread Spectrum
FTP-File Transfer Protocol
GP-Guard Period
GPS-Global Positioning System
GSM-Global System For Mobile Communications
HARQ-Hybrid Automatic Repeat Request
H-FDD-Half Frequency Division Duplex
HSDPA-High Speed Downlink Packet Access
HTTP-Hypertext Transfer Protocol
IE-Information Elements
IFFT-Inverse Fast Fourier Transform
IP-Internet Protocol
ISI-Inter Symbol Interference
KEK-Key Encryption Key
LAN-Local Area Network
LinkID-Link Identifier
LLC-Logical Link Control
LMDS-Local Multichannel Distribution Service
LMSC-LAN/MAN Standards Committee
LOS-Line Of Sight
LSB-Least Significant Bit
MAC-Medium Access Control
MAN-Metropolitan Area Network
MIB-Management Information Base
MIMO-Multiple In Multiple Out
MMDS-Multichannel Multipoint Distribution Service
MPDU-MAC Protocol Data Unit
MPLS-MultiProtocol Label Switching
MRC-Maximal Ratio Combining
MSB-Most Significant Bit

MSDU-MAC Service Data Unit
MS-Mobile Station
NACK-Negative Acknowledgement
NLOS-Not Line of Sight
NodeID-Node Identifier
NRTPS-Non-Real Time Polling Service
OFDMA-Orthogonal Frequency Division Multiple Access
OFDM-Optical Frequency Division Multiplexing
OFDM-Orthogonal Frequency Division Multiplexing
OSI-Open Systems Interconnection
PAN-Personal Area Network
PAPR-Peak to Average Power Ratio
PCF-Point Coordination Function
PCS-Packet Convergence Sublayer
PDU-Protocol Data Unit
PER-Packet Error Rate
PHSI-Payload Header Suppression Index
PHSM-Payload Header Suppression Mask
PHS-Payload Header Suppression
PHY-Physical Layer
PICS-Protocol Implementation Conformance Statement
PKM-Privacy Key Management Protocol
PM-Poll Me Bit
PMP-Point to Multipoint
PN-Packet Number
PRBS-Pseudo-Random Binary Sequence
PSH-Packing Subheaders
PS-Physical Slot
PTP-Point to Point
QAM-Quadrature Amplitude Modulation
QoS-Quality Of Service
QPSK-Quadrature Phase Shift Keying
RCID-Reduced Connection Identifier
RCT-Radio Conformance Tests
RF-Radio Frequency
RLC-Radio Link Control Protocol
RSSI-Received Signal Strength Indicator
RSV-Reserved

RTG-Receive-Transmit Transition Gap
RTPS-Real Time Polling Service
RTS-Request To Send
SAID-Security Association Identifier
SAP-Service Access Point
SA-Security Association
SC-Single Carrier
SDMA-Spatial Division Multiple Access
SDO-Standards Development
Organizations
SDU-Service Data Unit
SFID-Service Flow Identifier
SHA-1-Secure Hash Algorithm Number 1
SI-Slip Indicator
SLA-Service Level Agreement
SNMP-Simple Network Management
Protocol
SNR-Signal To Noise Ratio
SPID-Subpacket Identifier
SR-Selective Repeat
SSRTG-Subscriber Station Receive-
Transmit Transition Gap
SS-Subscriber Station
SSTG-Subscriber Station Transition Gaps
SSTTG-Subscriber Station Transmit-
Receive Transition Gap
TCM-Trellis Coded Modulation
TCP-Transmission Control Protocol
TDD-Time Division Duplex
TDMA-Time Division Multiple Access
TDM-Time Division Multiplexing
TEK-Traffic Encryption Key
TFTP-Trivial File Transfer Protocol
TLV-Type Length Value
TOS-Type Of Service
TSS&TP-Test Suite Structure and Test
Purposes
TTG-Transmit-Receive Transition Gap
UBR-Unspecified Bit Rate
UCD-Uplink Channel Descriptor
UDP-User Datagram Protocol
UIUC-Uplink Interval Usage Code

UL-Uplink
VCI-Virtual Channel Identifier
VC-Virtual Channel
VLAN-Virtual Local Area Network
VoIP-Voice Over Internet Protocol
VPI-Virtual Path Identifier
VP-Virtual Path
WiMax-Worldwide Interoperability for
Microwave Access
WLAN-Wireless Local Area Network
WPAN-Wireless Personal Area Network

Index

Printed in the United Kingdom
by Lightning Source UK Ltd.
124362UK00001B/70/A